Springer Proceedings in Materials 64

Springer Proceedings in Materials publishes the latest research in Materials Science and Engineering presented at high standard academic conferences and scientific meetings. It provides a platform for researchers, professionals and students to present their scientific findings and stay up-to-date with the development in Materials Science and Engineering. The scope is multidisciplinary and ranges from fundamental to applied research, including, but not limited to:

- Structural Materials
- Metallic Materials
- Magnetic, Optical and Electronic Materials
- Ceramics, Glass, Composites, Natural Materials
- Biomaterials
- Nanotechnology
- Characterization and Evaluation of Materials
- Energy Materials
- Materials Processing

To submit a proposal or request further information, please contact one of the following Springer Publishing Editors according to your affiliation location:

European countries: **Mayra Castro** (mayra.castro@springer.com)

India, South Asia and Middle East: **Swati Meherishi** (swati.meherishi@springer.com)

South Korea: **Smith Chae** (smith.chae@springer.com)

Southeast Asia, Australia and New Zealand: **Ramesh Nath Premnath** (ramesh.premnath@springer.com)

The Americas: **Michael Luby** (michael.luby@springer.com)

China and all the other countries or regions, as well as topics in materials chemistry: **Maggie Guo** (maggie.guo@cn.springernature.com)

This book series is indexed in **SCOPUS** and **EI Compendex** database.

Andrei Victor Sandu · Petrica Vizureanu ·
Mohd Mustafa Al Bakri Abdullah · Marcin Nabialek ·
Mohd Remy Rozainy Mohd Arif Zainol · Ion Sandu
Editors

Selected Papers from the International Conference on Innovative Research

ICIR EUROINVENT 2024

International Conference on Innovative Research, ICIR EUROINVENT

Iasi Romania, June 6–7, 2024

Springer

Editors
Andrei Victor Sandu
Gheorghe Asachi Technical University of Iasi
Iaşi, Romania

Petrica Vizureanu
Gheorghe Asachi Technical University of Iasi
Iaşi, Romania

Mohd Mustafa Al Bakri Abdullah
Universiti Malaysia Perlis
Perlis, Malaysia

Marcin Nabialek
Czestochowa University of Technology
Częstochowa, Poland

Mohd Remy Rozainy Mohd Arif Zainol
Universiti Sains Malaysia
George Town, Malaysia

Ion Sandu
Alexandru Ioan Cuza University of Iasi
Iaşi, Romania

ISSN 2662-3161 ISSN 2662-317X (electronic)
Springer Proceedings in Materials
ISBN 978-3-031-81197-5 ISBN 978-3-031-81198-2 (eBook)
https://doi.org/10.1007/978-3-031-81198-2

This Springer imprint is published by the registered company Springer Nature Switzerland AG
The registered company address is: Gewerbestrasse 11, 6330 Cham, Switzerland

Preface

The current Springer Proceeding is a collection of peer-reviewed articles presented at International Conference on Innovative Research ICIR Euroinvent 2024 Conference. The event was held in Iaşi, Romania, from June 6th to 7th, 2024.

The organizers are the Romanian Inventors Forum; the Faculty of Materials Science and Engineering, "Gheorghe Asachi" Technical University of Iasi, Romania; ARHEOINVEST Platform, Alexandru Ioan Cuza University of Iasi; the Centre of Excellence in Geopolymer and Green Technology (CEGeoGTech), Universiti Malaysia Perlis (UniMAP); Department of Physics, Czestochowa University of Technology, Częstochowa, Poland; and the National Institute for Research and Development in Environmental Protection (INCPDM).

The ICIR Conference is organized under the auspices of EUROINVENT. This is a joint event promoting creativity in a European context by displaying the contributions of consecrated schools from higher education and academic research and also of individual inventors and researchers.

The EUROINVENT International Conference on Innovative Research (ICIR) serves as a premier platform for top-tier researchers, engineers, and scientists to showcase their latest findings in materials science and engineering.

The volume covers all aspects of materials science, from the synthesis and characterization of materials to procedures and technologies for materials engineering, as well as materials applications and their involvement in the life sciences.

All the papers were reviewed by at least two expert referees in their relevant topic disciplines, and only nine were accepted. The papers selected for the volume were chosen based on their quality and relevance to the conference. All articles were checked for plagiarism using software.

The conference was very dynamic, with many questions and responses from the participants. During the conference closing ceremony, the Scientific Board decided to present the Best Oral Presentation Award (500 EUR), in addition to two Best Poster Awards.

The editors hope that this volume will provide the reader with a broad overview of the latest advances in the field of materials science and engineering, and that it will serve as a valuable reference source for further research.

The editors would like to express their sincere appreciation and thanks to all the committee members of the ICIR 2024 for their tremendous efforts.

Finally, the editors would like to extend their sincerest gratitude to all the authors for their valuable contributions to this volume.

Iași, Romania — Andrei Victor Sandu
Iași, Romania — Petrica Vizureanu
Perlis, Malaysia — Mohd Mustafa Al Bakri Abdullah
Częstochowa, Poland — Marcin Nabialek
George Town, Malaysia — Mohd Remy Rozainy Mohd Arif Zainol
Iași, Romania — Ion Sandu

Contents

Integrating Artificial Intelligence in Nanomaterials Science: Pathways to Revolutionary Materials Discovery and Design. Ethics and Risks

Alexandru Silviu Goga[1] (✉)

[1] University of Transylvania, Brasov, Romania
alexandru.goga@unitbv.ro

Abstract. The interdisciplinary nexus of artificial intelligence (AI) and nanomaterials science heralds a paradigm shift in materials discovery and application. This paper provides a comprehensive exploration of how AI is revolutionizing the field of nanotechnology, with a particular focus on the synthesis, characterization, and deployment of nanomaterials. We begin by defining nanomaterials and elucidating their unique, scale-dependent properties that make them integral to advancements in sectors such as health, electronics, and environmental science. Advancements in machine learning (ML) algorithms have unlocked unprecedented capabilities in predictive analytics, enabling the design of nanomaterials with tailored properties. We delve into the specifics of how ML and deep learning are employed to correlate nanomaterials' structural attributes with their functional characteristics, thus facilitating a more nuanced understanding of structure-property relationships. Furthermore, the paper examines AI's role in enhancing the resolution and interpretative power of characterization techniques like electron microscopy and spectroscopy. The discussion extends to mathematical data-driven discovery, where AI not only accelerates the identification of novel nanomaterials but also suggests optimal pathways for their synthesis. While the promise of AI in nanomaterials science is immense, the paper also addresses the challenges and ethical considerations inherent in this burgeoning field. Limitations of current AI methodologies in fully capturing the quantum mechanical behaviors at the nanoscale are discussed, as are the potential risks and environmental impacts. The conclusion encapsulates the synergistic potential of AI and nanotechnology, emphasizing the need for sustainable and responsible innovation. The future directions point towards an AI-integrated approach that is poised to redefine the landscape of material science and engineering.

Keywords: Artificial intelligence · Nanotechnology · Ethics

A. V. Sandu et al. (Eds.): ICIR EUROINVENT 2024, SPM 64, pp. 1–11, 2025.
https://doi.org/10.1007/978-3-031-81198-2_1

1 Introduction

1.1 Background on Nanomaterials and Their Significance

Imagine a world where materials are so tiny, they dance between the realms of classical physics and quantum mechanics, flaunting properties that can revolutionize everything from medicine to electronics. Welcome to the enigmatic world of nanomaterials. These pint-sized powerhouses, typically less than 100 nanometers in size, exhibit unique characteristics that their bulk counterparts can only dream of. For instance, gold, which is famously inert in bulk form, becomes a super-reactive catalyst at the nanoscale. It's like discovering your mild-mannered librarian moonlights as a rock star. The ability to manipulate matter at such a minute level opens up a Pandora's box of possibilities—enhanced drug delivery systems, ultra-strong yet lightweight materials, and highly efficient energy storage devices. Indeed, nanomaterials are the unsung heroes poised to tackle some of the most daunting challenges of our time.

1.2 Introduction to AI and Its Relevance to Nanomaterials Science

Now, let's add another layer of complexity and intrigue to the mix: artificial intelligence (AI) [1–4]. If nanomaterials are the superheroes of the material world, AI is the genius sidekick with a knack for solving problems that leave mere mortals scratching their heads. AI's prowess lies in its ability to process and analyze vast datasets with the speed and precision of a caffeine-fueled mathematician. In the realm of nanomaterials science, AI is not just an accessory; it's a game-changer. By deploying machine learning algorithms, scientists can predict the properties of new nanomaterials before they're even synthesized, saving both time and resources [5–8]. It's like having a crystal ball that not only shows you the future but also helps you create it. AI can also enhance the capabilities of characterization techniques, making it possible to visualize and understand the complex structures and behaviors of nanomaterials with unprecedented clarity. In essence, AI and nanomaterials together form a dynamic duo with the potential to transform industries and improve lives in ways we are only beginning to fathom [10–12].

1.3 Objectives and Scope of the Paper

So, what's the grand plan for this paper? Our mission, should we choose to accept it, is to explore the burgeoning intersection of AI and nanomaterials science. We'll embark on a journey through the synthesis, characterization, and application of nanomaterials, guided by the predictive and analytical prowess of AI. We aim to illustrate how AI is not just a tool but a transformative force that can unlock new dimensions in materials discovery and design. Along the way, we'll shine a light on the ethical considerations and potential risks, because with great power comes great responsibility. By the end of this exploration, we hope to provide a comprehensive understanding of how AI and nanomaterials can synergize to create a future where materials are not only smarter but also more sustainable. Buckle up; it's going to be a thrilling ride!

2 Nanomaterials: Definition and Properties

2.1 Definition and Types of Nanomaterials

Imagine being able to hold a whole new universe in the palm of your hand—welcome to the world of nanomaterials. These materials, defined by their minuscule size (typically less than 100 nanometers), pack an enormous punch in terms of properties and applications. To put their size into perspective, a single nanometer is a billionth of a meter. If a nanomaterial were scaled up to the size of a soccer ball, then a regular soccer ball would be as large as the Earth. There are various types of nanomaterials, each with its own unique set of capabilities.

These include nanoparticles, nanowires, nanorods, and nanotubes. Nanoparticles are the Swiss army knives of the nano world, used in everything from sunscreens to cancer treatment. Nanowires and nanorods, with their elongated structures, are excellent conductors and find applications in electronic devices. Meanwhile, carbon nanotubes, with their incredible strength and electrical conductivity, are like the superheroes of nanomaterials, promising to revolutionize everything from materials science to aerospace engineering.

2.2 Scale-Dependent Properties of Nanomaterials

Size matters, especially when you're talking about nanomaterials. At the nanoscale, materials exhibit properties that are vastly different from their bulk counterparts, thanks to the increased surface area and quantum effects. Imagine gold, the epitome of stability and bling, suddenly becoming a highly reactive catalyst at the nanoscale. It's like discovering that your demure pet cat has a hidden talent for breakdancing. The surface area to volume ratio increases dramatically as materials shrink, which enhances reactivity and catalytic efficiency.

Additionally, quantum effects start to dominate, leading to unique optical, electrical, and magnetic properties. For instance, nanoparticles can absorb and emit light in ways that bulk materials cannot, making them perfect for applications in imaging and diagnostics. The scale-dependent properties of nanomaterials are not just a scientific curiosity; they are the key to unlocking new technologies and solutions across various fields.

2.3 Applications in Health, Electronics, and Environmental Science

The versatility of nanomaterials opens doors to a plethora of applications that touch almost every aspect of our lives. In health, nanomaterials are the secret agents working behind the scenes. They enable targeted drug delivery systems that can seek out and destroy cancer cells with precision, minimizing damage to healthy tissues. Imagine a tiny, intelligent missile honing in on a tumor and obliterating it without harming the surrounding area. In the realm of electronics, nanomaterials are paving the way for faster, smaller, and more efficient devices. Think of smartphones with batteries that last for days or computers with processing power that leaves today's technology in the dust. Environmental science also reaps the benefits, with nanomaterials helping to clean up pollutants and create more efficient renewable energy solutions. Picture nanoparticles acting as sponges, soaking up oil spills, or as catalysts in solar cells, capturing sunlight with unmatched efficiency. From medicine to electronics to the environment, nanomaterials are not just the building blocks of the future—they are the architects of a smarter, cleaner, and more sustainable world.

3 Advancements in Machine Learning for Nanomaterials

3.1 Overview of ML and Deep Learning in Materials Science

In the bustling bazaar of materials science, machine learning (ML) and deep learning are the savvy traders who never sleep, constantly analyzing patterns and making predictions that would take human researcher's lifetimes to decipher. Imagine an algorithm with the curiosity of Sherlock Holmes and the computational brawn of a supercomputer. ML and deep learning in materials science are revolutionizing how we approach the discovery and design of new materials. By sifting through colossal datasets, these intelligent systems can identify trends and relationships that are otherwise invisible to the naked eye. Whether it's predicting the crystal structure of a new alloy or optimizing the synthesis of a complex polymer, ML and deep learning are the ultimate research assistants, tirelessly working behind the scenes to push the boundaries of what's possible in materials science [12–16].

3.2 Predictive Analytics and Design of Nanomaterials with Tailored Properties

Think of predictive analytics as the crystal ball of the twenty-first century—only, instead of vague prophecies, it offers precise, actionable insights. In the realm of nanomaterials, predictive analytics enables scientists to design materials with properties as specific as a bespoke suit. Want a nanoparticle that's super-efficient at converting solar energy? Or perhaps a nanofiber that's incredibly strong yet flexible? Predictive analytics can take the guesswork out of the equation. By using ML models trained on vast amounts of experimental data, researchers can predict how changes at the atomic level will affect macroscopic properties. This means we can design nanomaterials with tailored properties before even stepping into the lab, saving both time and resources. It's like having a recipe book where every dish is guaranteed to be a hit.

3.3 Structure-Property Relationships Facilitated by ML Algorithms

The relationship between a material's structure and its properties is like a complex dance—every twist and turn can dramatically affect the outcome. Traditionally, deciphering these structure-property relationships has been akin to solving a particularly fiendish puzzle. Enter ML algorithms, the puzzle masters who can untangle even the knottiest problems. By analyzing structural data from myriad sources, ML algorithms can uncover the intricate relationships between a nanomaterial's structure and its properties. This not only enhances our understanding of existing materials but also guides the creation of new ones. For instance, an ML model might reveal that a slight tweak in the arrangement of atoms in a nanomaterial could significantly boost its electrical conductivity. It's like discovering that a minor adjustment in a dance step can transform a good performance into a showstopper. With ML algorithms facilitating these insights, the path from discovery to application becomes much more efficient, paving the way for innovations that were previously beyond our reach.

4 AI in Characterization Techniques

4.1 Enhancements in Resolution and Interpretative Power

In the world of nanomaterials, seeing truly is believing, and when it comes to peering into the infinitesimal, the devil is in the details. AI steps in as the ultimate magnifying glass, enhancing both the resolution and interpretative power of characterization techniques. Think of it as trading in your old reading glasses for a state-of-the-art microscope that can not only see smaller but also understand what it's looking at. Traditional imaging methods can struggle with the sheer complexity and scale of nanomaterials, but AI algorithms excel at sifting through the noise to reveal hidden patterns and features. By improving resolution, AI allows scientists to visualize nanostructures with unprecedented clarity. But the real magic happens in the interpretative phase: AI doesn't just show you a clearer picture; it tells you the story behind it, identifying key features and anomalies that might otherwise go unnoticed.

4.2 AI's Role in Electron Microscopy and Spectroscopy

Electron microscopy and spectroscopy are like the Sherlock Holmes and Dr. Watson of nanomaterials characterization—one provides the keen eye, the other the detailed analysis. Now, imagine this dynamic duo being upgraded with the deductive power of an AI super-sleuth. In electron microscopy, AI enhances image resolution and sharpness, transforming blurry blobs into crisp, detailed structures. It's like adjusting the focus on a camera, but with an intelligence that can also identify and classify what it sees. Meanwhile, in spectroscopy, AI algorithms analyze the spectral data with lightning speed and accuracy, identifying material compositions and properties with a level of detail that's almost superhuman. AI can process vast amounts of data in real-time, spotting subtle shifts and trends that might elude even the most experienced human analyst. It's as if Holmes and Watson just got their hands on a high-tech gadget that not only magnifies their observations but also interprets them with impeccable precision.

4.3 Case Studies Illustrating AI's Impact on Characterization

Consider the case of a team of researchers trying to characterize a new type of carbon nanotube. Traditional methods would have them laboriously analyzing electron microscopy images and spectroscopy data, a process akin to assembling a jigsaw puzzle in the dark. Enter AI, stage left. By applying deep learning algorithms, the team can rapidly process and interpret their data, identifying defects and variations in the nanotubes with pinpoint accuracy. One real-world example saw AI-enabled electron microscopy reveal atomic-scale defects in graphene sheets that were previously invisible, providing crucial insights into how these defects affect material properties. Another case involved AI-enhanced spectroscopy unraveling the complex composition of multi-layered nanomaterials, allowing for precise tuning of their electronic properties. These case studies demonstrate that AI is not just a tool but a transformative force in characterization, turning what was once a painstaking process into an efficient, insightful journey into the nanoworld.

5 Mathematical Approach: AI in Nanomaterial Innovation

A mathematical equation to represent the optimization of research in nanomaterials science can incorporate various factors, including material properties, experimental parameters, and computational models. One simple way to express this is through an objective function that we aim to maximize or minimize, depending on the goal. Here's an example of such an equation:

$$\sigma = f(M, E, C)$$

where:

- σ is the objective function representing the desired outcome (e.g., efficiency, strength, cost-effectiveness).
- M is a vector of material properties (e.g., conductivity, tensile strength, thermal stability).
- E is a vector of experimental parameters (e.g., temperature, pressure, time).
- C is a vector of computational model parameters (e.g., algorithm choice, computational power, accuracy).

To make it more concrete, let's consider optimizing the electrical conductivity (σ\sigmaσ) of a nanomaterial.

The equation might look like:

$$\text{Maximize}\,\sigma = f\left(M1, M2, E1, E2, C1, C2\right)$$

where:

- M1 and M2 are specific material properties like atomic composition and nanostructure.
- E1 and E2 are experimental conditions like synthesis temperature and time.
- C1 and C2 are computational parameters like simulation accuracy and model type.

In a practical optimization scenario, this could be solved using techniques such as gradient descent, genetic algorithms, or machine learning models to find the best combination of M, E, C that maximizes the electrical conductivity σ.

5.1 Data-Driven Discovery and Identification of Novel Nanomaterials

Picture this: a vast ocean of data, filled with countless possible combinations and configurations of atoms. Traditional methods of material discovery are akin to fishing with a single hook, slow and often fruitless. Enter data-driven discovery, the industrial-sized trawler equipped with AI algorithms, capable of analyzing massive datasets with unprecedented speed and accuracy. By leveraging big data and machine learning, researchers can sift through this ocean, identifying novel nanomaterials with unique properties that were previously hidden beneath the surface. It's like finding pearls where others saw only sand. One striking example is the use of AI to predict new high-entropy alloys, a class of materials with superior strength and thermal stability. By analyzing vast compositional spaces, AI has identified combinations that human researchers might never have considered, paving the way for materials with groundbreaking applications in aerospace and beyond.

5.2 AI-Suggested Pathways for Nanomaterial Synthesis

Imagine having a GPS for nanomaterial synthesis, guiding you through the most efficient and effective routes to your destination. That's precisely what AI-suggested pathways offer to researchers. By analyzing previous synthesis attempts, experimental data, and theoretical models, AI can propose optimal pathways for creating new nanomaterials. It's like having an experienced chef whispering the perfect recipe in your ear, complete with tips on avoiding common pitfalls. These AI-driven suggestions are not just educated guesses; they are based on a deep understanding of chemical reactions, thermodynamics, and kinetics. For instance, AI has been used to optimize the synthesis of perovskite solar cells, suggesting conditions that maximize efficiency and stability. The result? Faster development cycles, lower costs, and materials that push the boundaries of what's technologically possible. With AI-suggested pathways, the journey from concept to reality becomes not just shorter but also more predictable and successful.

6 Ethical Considerations and Risks

6.1 Limitations of Current AI Methodologies

AI in nanomaterials science is like having a high-tech Swiss army knife; incredibly versatile but not without its limitations. Despite the impressive capabilities of machine learning and deep learning, these algorithms sometimes resemble overenthusiastic interns—they work tirelessly but can occasionally get things hilariously wrong. Current AI methodologies can struggle with the sheer complexity and variability of nanomaterials [9]. Their predictions can be biased by the quality of data they are trained on, and let's face it, garbage in means garbage out. Moreover, these models often operate as black boxes, making it difficult to understand how they arrive at specific conclusions. This opacity can be a significant hurdle in scientific research, where understanding the 'why' is as crucial as the 'what'. Until we develop more transparent and robust AI systems, these limitations will remain a stumbling block on the path to fully harnessing AI's potential in nanomaterials science.

6.2 Challenges in Capturing Quantum Mechanical Behaviors at the Nanoscale

When it comes to the nanoscale, even the most sophisticated AI can feel like a fish out of water. The quantum mechanical behaviors that dominate this tiny realm are notoriously tricky to capture. Think of trying to predict the weather on Mars while wearing a blindfold—there's a lot you're likely to miss. Quantum mechanics introduces a level of complexity that challenges our best models and algorithms, with phenomena such as quantum tunneling and entanglement defying classical logic. AI systems need to incorporate quantum computing and advanced theoretical models to make accurate predictions at this scale, but we're still in the early days of this integration. Until we crack this nut, capturing the full essence of quantum behaviors in nanomaterials will remain a formidable challenge.

6.3 Potential Environmental Impacts and Ethical Concerns

AI-driven advancements in nanomaterials hold immense promise, but they also come with a Pandora's box of potential environmental impacts and ethical concerns. Imagine creating the perfect material for a specific application, only to find it's as environmentally friendly as a coal-powered submarine. The production and disposal of nanomaterials can pose significant environmental risks, including toxicity and persistence in ecosystems. Moreover, the use of AI in this field raises ethical questions about data privacy, algorithmic bias, and the potential for job displacement. As we charge forward with these innovations, it's crucial to ask ourselves: Are we considering the full lifecycle of these materials? Are we ensuring that the benefits are equitably distributed?

Addressing these concerns requires a proactive approach, combining rigorous environmental assessments with ethical oversight to ensure that we are not just innovating but doing so responsibly.

6.4 Discussion on Responsible AI Use and Sustainable Innovation

Responsible AI use and sustainable innovation should be our guiding stars as we navigate the uncharted waters of nanomaterials science. Picture AI as a powerful engine capable of driving remarkable progress—but without a skilled driver and a clear roadmap, it can easily veer off course. To harness this potential responsibly, we must establish robust ethical guidelines and ensure transparency in AI decision-making processes. This involves not just technologists but also ethicists, policymakers, and the public in the conversation. Sustainable innovation, meanwhile, calls for a holistic approach, considering environmental, social, and economic impacts at every stage of development. By prioritizing renewable resources, minimizing waste, and ensuring that new technologies benefit all of society, we can build a future where nanomaterials and AI not only coexist but thrive together. In the end, it's about striking the right balance—leveraging AI's immense capabilities while keeping our feet firmly planted on the ground of ethical responsibility and sustainability.

7 Future Directions and Conclusion

7.1 Synergistic Potential of AI and Nanotechnology

In the grand theater of scientific innovation, AI and nanotechnology are the leading actors in a performance that promises to captivate the world. Their synergistic potential is nothing short of revolutionary, akin to combining the intellectual genius of Sherlock Holmes with the technological prowess of Iron Man. AI's ability to analyze and predict, when paired with nanotechnology's capacity to manipulate matter at the atomic level, creates a powerhouse duo that can redefine the boundaries of what's possible. Imagine developing materials that self-repair, or sensors so sensitive they can detect single molecules of a toxin. This isn't science fiction—it's the imminent reality ushered in by the convergence of AI and nanotechnology. Together, they can solve complex problems faster and more efficiently, leading to breakthroughs that were previously unimaginable.

7.2 Importance of Sustainable and Responsible Innovation

However, with great power comes great responsibility. As we stand on the cusp of this new era, the importance of sustainable and responsible innovation cannot be overstated. It's not enough to simply forge ahead with technological advancements; we must also consider their long-term impacts on our planet and society. Think of it as conducting a symphony where every note must harmonize perfectly—not just for the immediate audience but for future generations. This means rigorously assessing the environmental footprint of new nanomaterials, ensuring ethical AI deployment, and striving for inclusivity in the benefits these technologies offer. By embedding sustainability and responsibility into the very fabric of our innovation processes, we can ensure that progress does not come at the expense of our ecological or moral integrity.

7.3 Future Research Directions and Potential Advancements in Material Science and Engineering

Looking to the future, the horizon of material science and engineering gleams with potential advancements that are both exciting and profound. Future research directions are likely to focus on enhancing the integration of quantum computing with AI, to better understand and exploit quantum phenomena in nanomaterials. Additionally, the development of more sophisticated and transparent AI models will be crucial in demystifying the 'black box' and fostering trust in AI-driven discoveries. We can also anticipate the rise of smart materials—nanomaterials that can adapt and respond to their environment, opening up new frontiers in everything from medical implants to adaptive building materials. It's a thrilling vista of innovation, where each step forward is a leap towards a future that blends technological brilliance with sustainable and ethical foresight. The possibilities are vast, and as we continue to explore and innovate, we can be confident that the best is yet to come.

In conclusion, the convergence of artificial intelligence (AI) and nanotechnology represents a pivotal moment in the trajectory of material science and engineering. This synergy holds transformative potential, akin to combining the precision of a master craftsman with the computational prowess of a supercomputer. AI's ability to analyze vast datasets and predict complex phenomena, when applied to nanotechnology, empowers researchers to design and manipulate materials at scales previously unimaginable.

However, with this great potential comes the responsibility to ensure that innovation is not pursued at the expense of ethical considerations and environmental sustainability. As we forge ahead into this new era, it is imperative to prioritize sustainable practices and responsible innovation. This means developing nanomaterials that not only perform optimally but also minimize environmental impact throughout their lifecycle—from production to disposal. Moreover, ethical guidelines must be established to govern the use of AI in material science, addressing issues such as data privacy, algorithmic bias, and societal impact.

Looking forward, the future of material science and engineering lies in an AI-integrated approach that marries technological advancement with ethical awareness. This approach promises to redefine how we perceive and utilize materials, paving the way for innovations that enhance human life while preserving our planet. By embracing sustainable and responsible innovation, we can harness the full potential of AI and nanotechnology to create a future where materials are not just functional, but also ethical and environmentally sound. Together, let us embark on this journey towards a brighter and more sustainable future, where science and technology harmonize to benefit all of humanity.

References

1. Eiben, A.E., Smith, J.E.: Introduction to Evolutionary Computing. Springer, Berlin, Heidelberg (2003)
2. Erik, J.L.: Myth of AI. Polirom, Bucharest (2022)
3. Feraru, A., Ceocea, C., Turcu, O.L., Popovici, N.: Knowledge-Based Management Particularities in Small and Medium-Sized Enterprises in the North-Eastern Region of Romania. Trivent Publishing © Trivent, Arad (2020)
4. Kelly, K.: Inevitabile, the 12 Techs that Shape Our Future. Sapiens, Cluj-Napoca (2023)
5. Mircea, B., Costel, C., Alexandru-Silviu, G.: The Advent of Artificial Intelligence. Lambert Academic Publishing, Bucharest (2023)
6. Taleb, N.: The Black Swan: The Impact of the Highly Improbable. Random House, New York (2007)
7. Russell, S., Norvig, P.: Artificial Intelligence: A Modern Approach, 4th edn. Prentice Hall, Hoboken, NJ (2021)
8. Walsh, T.: 2062: A World Created by AI. Rao, Bucharest (2021)
9. Victor, N.: Rețele Neurale pentru Explorarea Datelor. Matrix Rom, Bucharest (2018)
10. Wucker, M.: The Gray Rhino: How to Recognize and Act on the Obvious Dangers We Ignore. Macmillan, London (2016)
11. Ashwani, K., Venkatesh, M., Vranda, J., Himanshu, G.: Managing healthcare supply chain through artificial intelligence (AI): a study of critical success factors. Comput. Ind. Eng. **175**(5), 108815 (2022)
12. Berger, K., Carter, A., Jones, M., Luttig, L., Smith, A., Watterson, B.: Scaling Gen AI in Banking: Choosing the Best Operating Model. https://www.ourperspectives.com/article/30492-scaling-gen-ai-in-banking-choosing-the-best-operating-model (2024)
13. Acharya, R., Keller, A., Klinger, I., Liu, A.: How Generative AI Can Help Banks Manage Risk and Compliance. 4th of March 2024. https://www.mckinsey.com/industries/financial-services/our-insights/scaling-gen-ai-in-banking-choosing-the-best-operating-model (2024)
14. Benaich, N.: State of AI Report 2023. https://www.mckinsey.com/capabilities/quantumblack/our-insights/the-state-of-ai (2023)
15. Singh, A.V., Varma, M., Laux, P., Choudhary, S., Datusalia, A.K., Gupta, N., Luch, A., Gandhi, A., Kulkarni, P., Nath, B.: Artificial intelligence and machine learning disciplines with the potential to improve the nanotoxicology and nanomedicine fields: a comprehensive review. Arch. Toxicol. **97**(4), 963–979 (2023)
16. Hamilton, S., Kingston, B.R.: Applying artificial intelligence and computational modeling to nanomedicine. Curr. Opin. Biotechnol. **85**, 103043 (2023)

Life Cycle Assessment of Geopolymers Obtained from Bulgarian Industrial Wastes

Nina Penkova[1], Andriana Surleva[1], Kalin Krumov[1], Darya Ilieva[1], Petrică Vizureanu[2], and Dumitru Doru Burduhos-Nergiş[2](✉)

[1] University of Chemical Technology and Metallurgy, Sofia, Bulgaria
[2] Technical University of Iasi "Gheorghe Asachi", Iasi, Romania
bunduc.doru@yahoo.com

Abstract. A computational algorithm for material and energy balance of a technology for production of geopolymers from fly ash at coal combustion, copper mine tailings and sodium silicate solution is developed. It is applied for determination of utilized wastes, embodied energy and carbon footprint of the products. The initial, critical and equilibrium moisture content of the geopolymers, the drying shrinkage and the energy consumption of the process in industrial conditions are obtained by experimental drying. For energy saving, options for free drying in environmental conditions are discussed. The influence of the shrinkage during the casting of the geopolymer bodies on their design and use in the construction is considered.

Keywords: Geopolymers · Waste utilization technology · Fly ash · Mine tailing · Material balance · Drying · Shrinkage · Energy consumption

1 Introduction

The utilisation of industrial wastes to produce building materials is a sustainable practice for resource and energy saving with ecological footprint. The geopolymers manufacturing is such an approach, provoking scientific and educational interests [1–3]. Many Bulgarian raw materials as fly ash, metakaolin, natural zeolite, iron-rich wastes from copper industry etc. are suitable for production of geopolymers [4].

It is established that geopolymers with sufficient strength to be used as building materials can be produced by Bulgarian copper mine tailing, fly ach and sodium silicate solution [5]. An international team developed such a technology during the implementation of a Project "RecMine – Environmental footprint reduction through eco-friendly technologies of mine tailings recycling", funded by ERA-MIN3 program of European Union. The aim of the present study is to specify the energy intensive technological processes and to work out computational algorithms for assessment of the embodied raw materials, energy and carbon dioxide at future industrial applications of the technology.

A. V. Sandu et al. (Eds.): ICIR EUROINVENT 2024, SPM 64, pp. 13–24, 2025.
https://doi.org/10.1007/978-3-031-81198-2_2

2 Technology Description

A common scheme of the proposed technology is presented on Fig. 1. The used raw materials for the geopolymer manufacturing present coal combustion fly ash (FA), cooper mine tailings (MT), sodium silicate solution (water glass), sodium hydroxide flakes (NaOH) and water.

The detailed procedure of geopolymer obtaining is presented in Vizureanu and co-workers [5]. The solid components FA and MT are industrial wastes in bulk. They are mixed in dry state for about 3 minutes to obtain homogeneous mixture. The materials were dried beforehand at the laboratory experiments. However, the drying could be avoided: the water content of the raw materials is relatively low and will slow down the curing process insignificantly.

The liquid component is obtained by mixing the sodium hydroxide solution with the sodium silicate solution. The NaOH solution is prepared by dissolving the NaOH flakes in water at least 24 hours before the mixing the solid and liquid to eliminate the influence of the exothermic NaOH dissolution reaction on the geopolymerization process.

The solid and liquid are mixed until a homogeneous slurry is obtained. The slurry is poured into molds and shaken for about 1 hour to allow the air to leave the liquid mass.

The specimens are left in the molds for 24 hours after the casting for initial solidification. The castings are then removed and allowed to further hardening and drying in the ambient environment (free drying) or in industrial dryers (forced drying).

During the initial experiments the casting are wrapped in foil to prevent them from drying out quickly. When conducting experiments to study the drying shrinkage and kinetics of the molded specimens, it is found that the drying rate of the molded specimens is low compared to the drying of construction ceramic products under equal conditions. This allows not using foil covers to slow down the process.

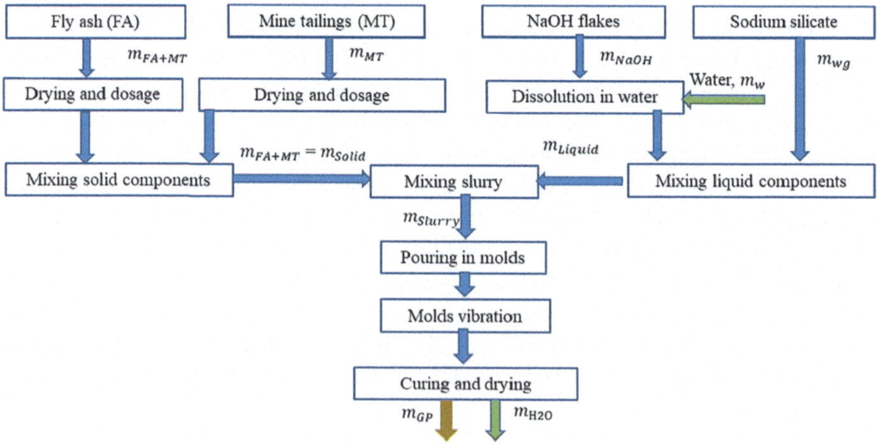

Fig. 1. Flow diagram for manufacturing of geopolymers.

Table 1. Compositions of the solid components.

Oxide	FA	MT	Dry mixture FA + MT
-	wt.%	wt.%	wt.%
SiO_2	41.2	65.5	47.28
Fe_2O_3	21.8	2.67	17.02
Al_2O_3	15.9	19.2	16.73
CaO	12.3	1.33	9.56
SO_3	3.5	3.13	3.41
K_2O	2.7	4.7	3.20
TiO_2	1.2	0.32	0.98
Na_2O	0.67	0.97	0.75
BaO	0.34	0	0.26
SrO	0.182	0.032	0.14
ZrO_2	0.062	0.0105	0.05
MgO	0	1.87	0.47
P_2O_5	0	0.19	0.05
Other	0.05	0.05	0.05
Total	99.90	99.97	99.92

Appropriate parameters of raw materials composition that yield the greatest strength of the products were established through planned experiments [5]:

FA: MT ratio, wt.%: = 75: 25;
Liquid to solid mass ratio = 0.7;
Na: Al mass ratio = 1.

The solid component includes FA, MT and NaOH flakes. The liquid is consisted of water and sodium silicate solution.

The composition of the mixed FA ad MT is presented in Table 1. As a result of thermogravimetric analysis of the raw materials, it was found that the water losses on during a 2 hour hold at 120 degrees, are 0.20% at MT and 0.15% at FA. Therefore, the total moisture content of the raw materials is below 0.2% and can be neglected in the present analysis.

The masses of the raw materials are obtained via the material balance below.

3 Material Balance

The wet slurry, obtained at the mixing of the liquid and solid components of the raw materials is:

$$m_{slurry} = m_{solid} + m_{liguid} = m_{FA+MT} + m_{NaOH} + m_w + m_{wg}, \text{ kg} \tag{1}$$

where m_{FA+MT}, m_{NaOH}, m_w and m_{wg} are the masses of the FA and MT mixture, NaOH flakes, water for NaOH solution and water glass (sodium silicate solution), respectively.

According to the material balance of the simultaneously curing and drying process:

$$m_{slurry} = m_{gp} + m_{H_2O}, \text{kg} \tag{2}$$

where m_{gp} and m_{H_2O} are the masses of the products – geopolymers and the water, evaporated at the drying.

The mass of the evaporated water is [6]:

$$m_{H_2O} = \left(u_m - u_e\right) m_{gp}^{dry} = m_{gp} \frac{u_m - u_e}{1 + u_e}, \text{kg} \tag{3}$$

where u_m=moisture content of the slurry at the molding, referred to the dry mass, kg·kg^{-1}; u_e= equilibrium moisture content, kg.kg^{-1} and m_{gp}^{dry} = dry geopolymer mass, kg:

$$m_{gp,dry} = \frac{m_{gp}}{1 + u_e}, \text{kg} \tag{4}$$

Then:

$$m_{slurry} = m_{gp} \frac{1 + u_m}{1 + u_e} = m_{FA+MT} + m_{NaOH} + m_w + m_{wg}, \text{kg} \tag{5}$$

$$m_{gp} = m_{slurry} \frac{1 + u_e}{1 + u_m}, \text{kg} \tag{6}$$

The mass of NaOH flakes m_{NaOH} is obtained at an accepted molar ratio MR_{target} = SiO_2:Na_2O in the water glass. According to [2] it is optimal in the range MR_{target} є (1.7 ÷ 1.9). Taking into account that the molar and the mass ratios of these oxides are approximatively equal, the mass x of Na_2O to be added is obtained by:

$$MR_{target} = \frac{m_{SiO_2}}{m_{Na2O} + x} \tag{7}$$

$$x = 0.01 * \left(\frac{SiO_2}{MR_{target}} - Na_2O \right) m_{w.g}, \text{kg} \tag{8}$$

where SiO_2 and Na_2O are the weight % of the oxides in the water glass. The mass of the NaOH is computed considering that 1 kmol Na_2O is added by 2 kmol NaOH:

$$m_{NaOH} = 0.01 \left(\frac{SiO_2}{MR_{target}} - Na_2O \right) m_{w.g} \frac{2\text{M}_{NaOH}}{\text{M}_{Na_2O}} = 0.0129 \left(\frac{SiO_2}{MR_{target}} - Na_2O \right) m_{w.g} \tag{9}$$

where M_{NaOH} and M_{Na_2O} are the correspondent molar masses, kg·kmol^{-1}.

The mass of the FA and MT are obtained at the accepted ratio between the masses of the Al and Na:

$$Weight\ ratio = \frac{m_{Na}}{m_{Al}} = 1 \quad (10)$$

So $m_{Na} = m_{Al}$ and

$$\frac{2 * M_{Na}}{M_{Na_2O}} \frac{Na_2O}{100} \cdot m_{wg} + \frac{M_{Na}}{M_{NaOH}} m_{NaOH} = \frac{2 * M_{Al}}{M_{Al_2O_3}} \frac{Al_2O_3}{100} \cdot m_{FA+MT} \quad (11)$$

$$m_{FA+MT} = \frac{\frac{M_{Na}}{M_{Al}} * \left(\frac{2}{M_{Na_2O}} \frac{Na_2O}{100} \cdot m_{wg} + \frac{m_{NaOH}}{M_{NaOH}} \right)}{\frac{2}{M_{Al_2O_3}} \frac{Al_2O_3}{100}}, \mathrm{kg} \quad (12)$$

where Al_2O_3 is the percentage of the oxide in the raw mixture (Table 1) and M_{Na}, M_{Al} and $M_{Al_2O_3}$ are molar masses, kg·kmol^{-1}.

The coupling of equations (9) and (12), gives:

$$m_{FA+MT} = \frac{M_{Na}}{M_{Al}} \frac{M_{Al_2O_3}}{M_{Na_2O}} \frac{SiO_2 * m_{w.g}}{MR_{target} * Al_2O_3} = 1.4 \frac{SiO_2\, m_{w.g}}{MR_{target}\, Al_2O_3} \mathrm{kg} \quad (13)$$

The masses of the fly ash and the tailing are:

$$m_{FA} = 1.4 * 0.75 \frac{SiO_2 m_{w.g}}{MR_{target} * Al_2O_3} = 1.05 \frac{SiO_2 m_{w.g}}{MR_{target} Al_2O_3}, \mathrm{kg} \quad (14)$$

$$m_{MT} = 1.4 * 0.25 \frac{SiO_2 m_{w.g}}{MR_{target} Al_2O_3} = 0.35 \frac{SiO_2 m_{w.g}}{MR_{target} Al_2O_3}, \mathrm{kg} \quad (15)$$

The amount of the water is obtained from the liquid to solid ratio:

$$\frac{m_{wg} + m_w}{m_{FA+MT} + m_{NaOH}} = \frac{Liquid}{Solid} \quad (16)$$

Therefore:

$$m_w = \frac{Liquid}{Solid} * \left(m_{FA+MT} + m_{NaOH} \right) - m_{wg}, \mathrm{kg} \quad (17)$$

Taking into account (9) and (13):

$$m_w = m_{wg}\left(\frac{Liquid}{Solid}\left(1.4\frac{SiO_2}{Al_2O_3}\frac{1}{MR_{target}}+0.0129\left(\frac{SiO_2}{MR_{target}}-Na_2O\right)\right)-1\right),\text{kg} \quad (18)$$

So:

$$m_{slurry} = m_{wg}\left(1.4\frac{SiO_2}{Al_2O_3}\frac{1}{MR_{target}}+0.0129\left(\frac{SiO_2}{MR_{target}}-Na_2O\right)\right)\left(\frac{Liquid}{Solid}+1\right) \quad (19)$$

Accepting MR_{target} =1.7 and $\frac{Liquid}{Solid}=0.7$:

$$m_{slurry} = m_{wg}\left(1.4\frac{SiO_2}{Al_2O_3}+0.0129\left(SiO_2-1.7Na_2O\right)\right),\text{kg} \quad (20)$$

Then:

$$m_{gp} = m_{wg}\left(1.4\frac{SiO_2}{Al_2O_3}+0.0129\left(SiO_2-1.7Na_2O\right)\right)\frac{1+u_e}{1+u_m},\text{kg} \quad (21)$$

The mass of the water glass is computed by the desired amount of the geopolymers:

$$m_{wg} = \frac{m_{gp}}{\left(1.4\frac{SiO_2}{Al_2O_3}+0.0129\left(SiO_2-1.7Na_2O\right)\right)}\frac{1+u_m}{1+u_e},\text{kg} \quad (22)$$

The masses of the raw materials for obtaining of 1 kg geopolymer (m_{gp} = 1 kg) are:

$$m_{wg} = \frac{1}{\left(1.4\frac{SiO_2}{Al_2O_3}+0.0129\left(SiO_2-1.7Na_2O\right)\right)}\frac{1+u_m}{1+u_e},\text{kgkg}^{-1} \quad (23)$$

$$m_{NaOH} = 0.0129\left(\frac{SiO_2}{1.7}-Na_2O\right)\frac{1}{\left(1.4\frac{SiO_2}{Al_2O_3}+0.0129\left(SiO_2-1.7Na_2O\right)\right)}\frac{1+u_m}{1+u_e},\ \text{kgkg}^{-1} \quad (24)$$

$$m_{FA+MT} = 0.824\frac{SiO_2}{Al_2O_3}\frac{1}{\left(1.4\frac{SiO_2}{Al_2O_3}+0.0129\left(SiO_2-1.7Na_2O\right)\right)}\frac{1+u_m}{1+u_e},\text{kgkg}^{-1} \quad (25)$$

Table 2. Raw materials.

Component	Mass, kg	Mass %	Materials consumption per 1 kg, $kgkg^{-1}$
Sodium hydroxide (NaOH)	0.006	2.2%	0.024
FA	0.114	42.5%	0.479
MT	0.038	14.2%	0.160
Water	0.011	4.1%	0.046
Water glass	0.100	37.1%	0.418
Total (slurry)	0.258	100.0%	1.127

$$m_w = 0.41\frac{1+u_m}{1+u_e}\left(1-\frac{1}{1.4\frac{SiO_2}{Al_2O_3}+0.0129\left(SiO_2-1.7Na_2O\right)}\right), kgkg^{-1} \quad (26)$$

All masses above are computed at a known water glass composition. The amounts of the raw materials for a sodium silicate solution, containing 13.7 wt. % Na_2O, 31 wt. % SiO_2 and accepted $m_{wg} = 1$ kg is shown on Table 2. The material consumptions per 1 kg geopolymers in the last column are computed via the experimentally obtained equilibrium and molding water contents of the slurry, as it is given in the next section of the paper.

4 Drying Kinetics of the Developed Geopolymers

The initial idea is to develop a technology based on free drying at environmental conditions for low embodied energy of the geopolymers. To obtain information about such a process 3 samples of 14 mm × 47 mm × 12 mm were formed from a slurry with the composition in Table 2, prepared according to the technology on Fig. 1.

They were dried at ambient conditions: temperature of 20 °C and relative air humidity RH = 60%. It was established that the drying process continues more than 30 days (Fig. 2). A shrinkage (relative strain) of about 3–4% is found (Fig. 3).

To obtain detail information about the drying kinetics forced drying is implemented at fixed laboratory conditions: temperature of 90 °C and relative air humidity RH ≈ 2%. Four samples are dried continuously 48 h to approximatively zero moisture content. The results are shown on Figs. 4, 5 and 6.

Three periods of drying are established [6]:

- Initial (Zero) period of drying: the moisture content decrease with relatively high drying rate from 15% ($u_m = 0.15$ kg·kg^{-1}) to about 11% ($u = 0.11$ kg·kg^{-1});
- Constant rate drying period (I Drying Period): the drying rate is almost constant, the moisture content decrease from average value of 11% to critical moisture content of 3% ($u_{cr} = 0.03$ kg·kg^{-1}) at which the shrinkage stops;

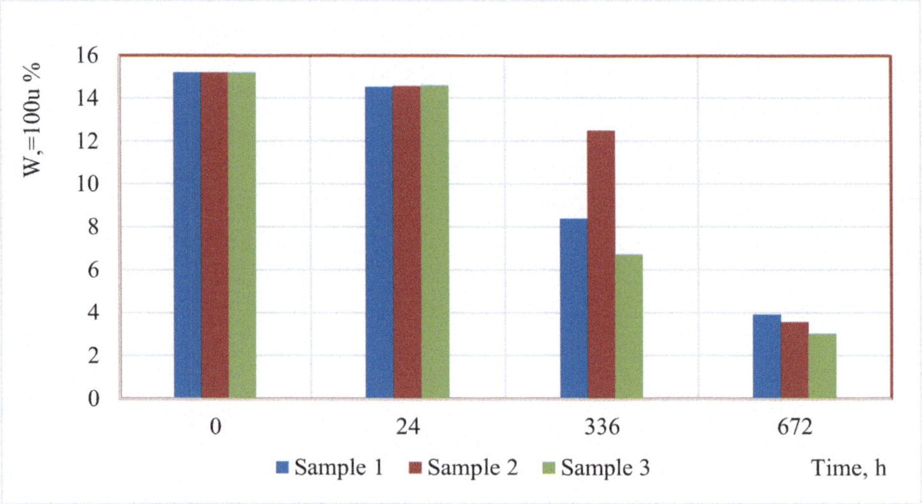

Fig. 2. Moisture content during drying at ambient conditions.

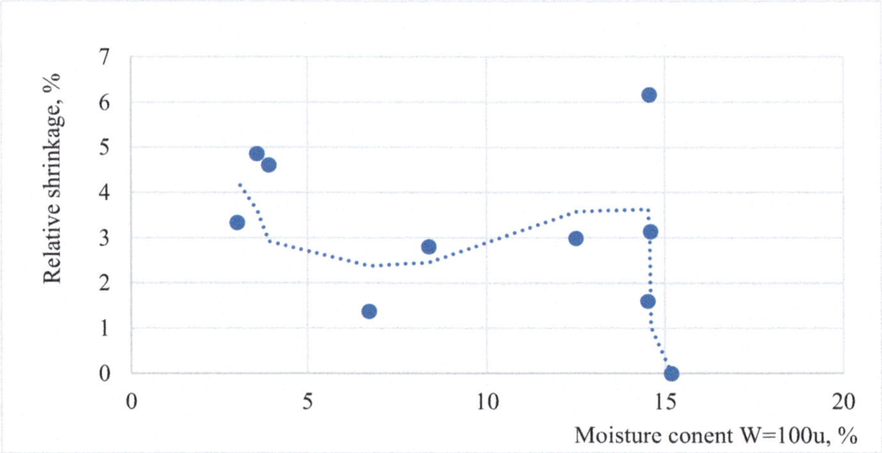

Fig. 3. Shrinkage of the material during a drying at ambient conditions.

- Falling rate drying period: a period with negligible shrinkage (lower than 0.5%). The moisture content is decreased form the critical to the equilibrium (at ambient condition) moisture content.

An equilibrium moisture content of $u_{eq} = 0.02\ kg \cdot kg^{-1}$ (2%) is computed based on the initial moisture content $u_{in} = 0.015\ kg \cdot kg^{-1}$ and the data, obtained at the free drying.

The experimentally forced drying at 90 °C and RH = 2% took relatively long time: the equilibrium moisture content was reached after 25 hours. Cracks and failures of the samples ware not occurred. Therefore, the temperature of 90 °C is suitable at industrial drying. However, the relative humidity of the drying media has to be relatively high at

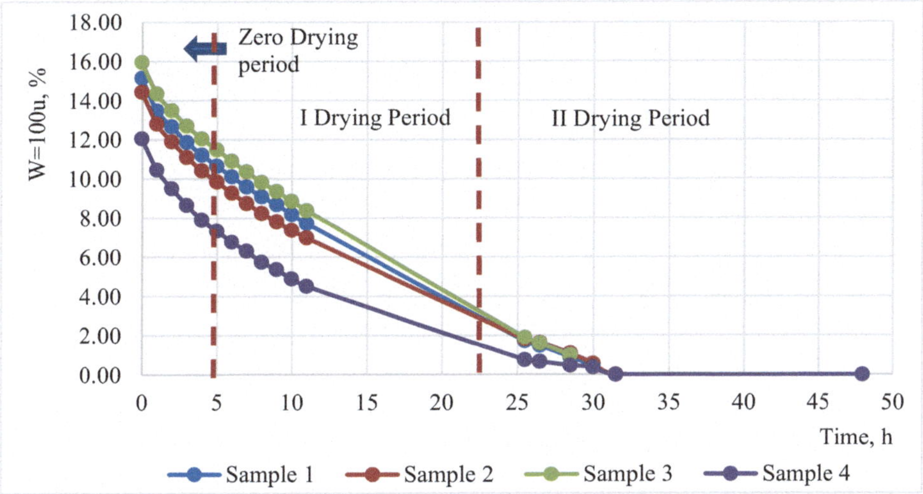

Fig. 4. Moisture content of geopolymer material at laboratory forced drying.

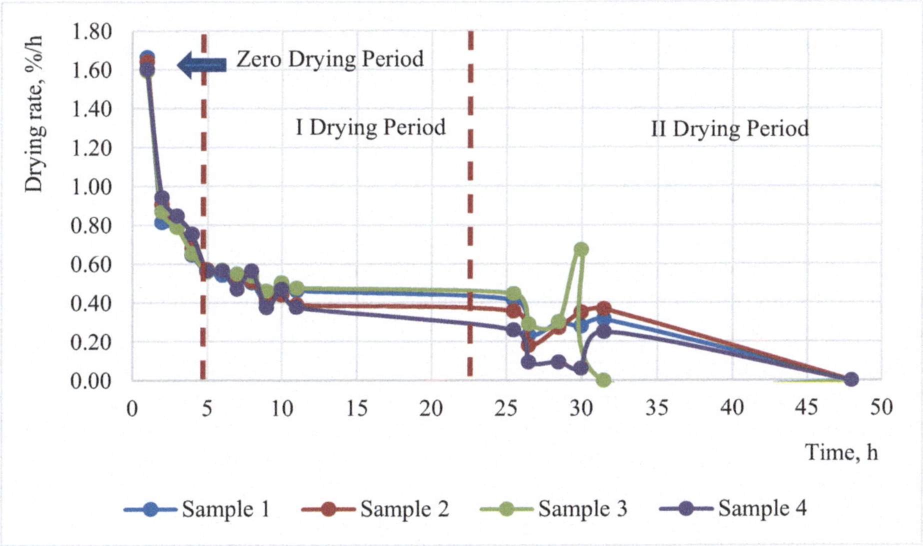

Fig. 5. Change of the drying rate at laboratory forced drying.

Zero and I period of the process to keep soft drying in order to prevent failure of the products, especially if their shape is massive according to the theory of heat transfer.

The evaporated water par 1 kg geopolymer material at the drying is computed according to eq. (3):

$$m_{\mathrm{H2O}} = m_{gp}\frac{u_m - u_e}{1 + u_e} = 1\frac{0.15 - 0.02}{1 + 0.02} = 0.127\,\mathrm{kg.kg}^{-1}$$

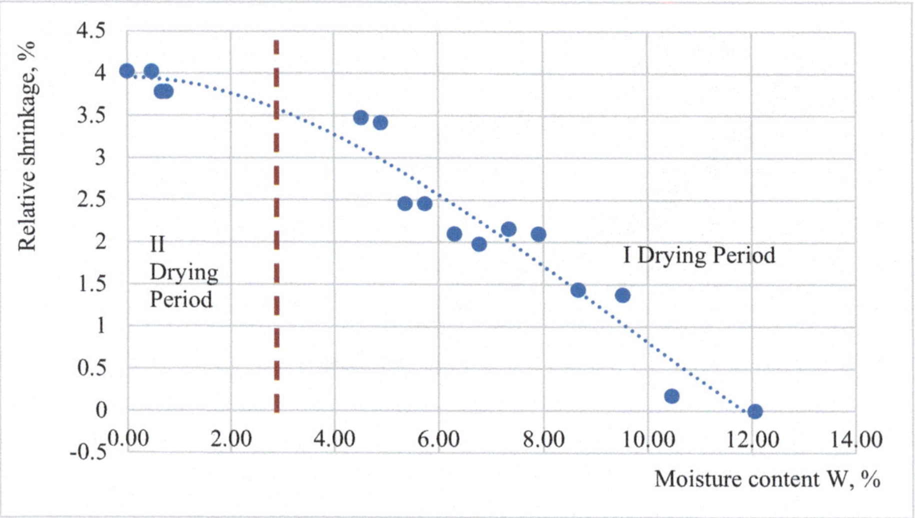

Fig. 6. Shrinkage of the material at laboratory forced drying.

It is necessary to take into account the drying shrinkage of 4% when determining of the initial dimensions of the molds at production of large-sized buildings elements, since this shrinkage cannot be compensated for, as in masonry.

5 Energy Consumption

The main energy consuming processes are the raw materials transportation, mixing and dosage, forming (molding) and drying. The energy input depends on the manufacturing level. For example, at small-scale plants, handmade molding and drying at ambient conditions, the energy consumption is expected to be negligible (electrical energy). The large-scale manufacturing with automatized machinery shaping and forces drying is energy intensive. Approximatively energy consumptions at the mechanized mixing and shaping are given in the next table. The energy inputs for the shaping, mixing and transportation are accepted according to [7]. The energy for drying is computed at accepted efficiency of continuous industrial dryer $\eta = 0.6$ (60%) according to [8]:

$$e = \frac{r.m_{\mathrm{H2O}}}{3600\eta}, \mathrm{kWh} \cdot \mathrm{kg}^{-1} \tag{27}$$

where r = evaporation energy per 1 kg of water: r = 2500 kJkg^{-1}.

The CO_2 emissions per 1 kg geopolymers in the last column in Table 3 are computed via the coefficients of equivalent emissions K_e of the energy resources according to the last Bulgarian regulation [9]:

- electrical energy: $K_e = 486$ g CO_2/kWh;
- natural gas (for drying): $K_e = 220$ g CO_2/kWh.

Table 3. Energy inputs and correspondent CO_2 emissions per 1 kg geopolymers.

Process	Energy consumption $MJ \cdot kg^{-1}$	Energy consumption $kWh \cdot kg^{-1}$	CO_2 emissions $kg \cdot kg^{-1}$
Weighing and Batch Preparation	0.060	0.017	0.008
Forming	0.160	0.044	0.022
Drying	0.529	0.147	0.032
Packaging	0.080	0.022	0.011
Total	0.829	0.230	0.073

Even at mechanised technological processes and industrial drying, the energy consumption to produce the developed geopolymers and the related carbon dioxide emissions are several times lower than the different ceramic products [10]. For example, the energy consumption of the masonry bricks is in the range 1.50–2.50 MJ/kg.

6 Conclusions

The production of geopolymers according to the proposed technology will lead to the utilization of industrial wastes: 0.479 kg fly ash and 0.160 kg mine tailings per kg product. This has a positive effect on the environment and on the consumption of natural raw materials for the construction sector.

The energy consumption and carbon dioxide emissions at industrial manufacturing of geopolymers according to the developed technology is expected to be at least 2 times lower compared to building ceramics. The energy inputs will be insignificant if the most energy-intensive process - drying – is carried out in the environmental conditions. However, it should be noted that this will increase the production time over a month.

The drying shrinkage of the material (4%) should be taken into account when forming the molds to obtain the desired geometry of the products.

Acknowledgments. Authors greatly acknowledge Project KP-06-DO 02/5 "RecMine – Environmental footprint reduction through eco-friendly technologies of mine tailings recycling", funded by Bulgarian Ministry of Education and Science for the support of this study.

References

1. Tanu, H.M., Unnikrishnan, S.: Utilization of industrial and agricultural waste materials for the development of geopolymer concrete – a review. Mater. Today Proc. **65**(2), 1290–1297 (2022)
2. Davidovits, J.: Geopolymer Chemistry and Applications, vol. 680, 5th edn. Institut Geopolymere, Saint-Quentin (2008)
3. Burduhos Nergis, D.D., Abdullah, M.M.A.B., Vizureanu, P., Tahir, M.F.M.: Geopolymers and their uses. In: IOP Conference Series: Materials Science and Engineering, p. 012019. IOP Publishing (2018)
4. Nikolov, A.: Geopolymers based on Bulgarian raw materials – preliminary studies. Int. Sci. J Mach. Technol. Mater. **13**(2), 197–199 (2019)

5. Ilieva, D., Angelova, L., Radoykova, T., Surleva, A., Chernev, G., Vizureanu, P., Burduhos-Nergis, D.D., Sandu, A.V.: Characterization of Bulgarian copper mine tailing as a precursor for obtaining geopolymers. Materials. **17**(3), 1996–1944 (2024)
6. Lykov, A.V.: Theory of Drying (in Russian), Moscow (1968)
7. Özkan, A., Günkay, Z., Tok, G., Karacasulu, L., Metesoy, M., Banar, M., Kara, A.: Life cycle assessment and life cycle cost analysis of magnesia spinel brick production. Sustain. For. **8**(7), 662 (2016)
8. Penkova, N.Y., Mladenov, B.M., Krumov, K.S.: Multiphysics simulation for design and adjustment of efficient industrial drying modes of ceramic ware. In: IOP Conference Series. Earth and Environmental Science, vol. 1128. IOP Publishing (2023)
9. Bulgarian regulation № РД-02-20-3/09.11.2022 for the technical requirements for energy characteristics of buildings (in Bulgarian). Bulgarian Ministry of the Energetics (2022)
10. Reference document on best available techniques in the ceramic manufacturing industry, European Commission (2007)

Fabrication of Very Fine ZnS Nanoparticles Through Surface Organo-Modification

Ionela-Mădălina Duinea[1,2], Aurel Tăbăcaru[1(✉)], Paul Chiriță[2], Nicoleta Cioateră[2], Alina-Mihaela Ceoromila[3], and Geta Cârâc[1(✉)]

[1] Department of Chemistry, Physics and Environment, Faculty of Sciences and Environment, "Dunărea de Jos" University of Galați, Galați, Romania
aurel.tabacaru@ugal.ro; geta.carac@ugal.ro

[2] Department of Chemistry, Faculty of Sciences, University of Craiova, Craiova, Romania

[3] Research and Development Center for Thermoset Matrix Composites, Cross-Border Faculty, "Dunărea de Jos" University of Galați, Galați, Romania

Abstract. Surface modification of inorganic nanomaterials through functionalization with organic capping agents may be regarded as a very useful and hand-on strategy to avoid particle agglomeration, thus improving the current properties and even to design new ones. In this context, we here report a simple, low-cost and high-yielding method for the fabrication of semiconductor zinc sulphide nanoparticles (ZnS NPs) by chemical co-precipitation, using the new capping agents tris(hydroxymethyl)aminomethane (TRIS) and 4-(2-hydroxyphenyl)-2-(morpholin-4-yl)-1,3-thiazole (DF). Powder X-ray diffraction (PXRD), scanning electron microscopy with energy dispersive X-ray spectroscopy (SEM-EDX), Fourier transform infrared spectroscopy (FTIR) and UV–visible spectroscopy allowed to confirm the important effect of the adopted capping agents in the preparation of ZnS NPs with size reduction from *ca.* 40 nm in unmodified ZnS NPs to *ca.* 10 nm in ZnS NPs modified with either TRIS or DF capping agents. Such an effect comes along with enhanced photoluminescence of the organo-modified ZnS NPs, with the maximum emission centred at 560 nm, and an increase of the band gap energy from 3.71 eV in unmodified ZnS NPs to 3.83 eV in organo-modified ZnS NPs.

Keywords: ZnS nanoparticles · Organo-modification · Optical properties

1 Introduction

Advanced research has recently been recorded for the synthesis and characterization of different zinc sulphide nanoparticles (ZnS NPs), with their special properties and applications in different fields of technological, industrial and environmental relevance [1–10]. The size feature has been shown to be a very important factor that offers ZnS NPs the advantage of formulating new composite materials with optimized

A. V. Sandu et al. (Eds.): ICIR EUROINVENT 2024, SPM 64, pp. 25–37, 2025.
https://doi.org/10.1007/978-3-031-81198-2_3

applications, such as optoelectronics, luminescence, photocatalysis, electrocatalysis, displays, sensors and absorption studies [4–10]. Among the various methods already known for the synthesis of ZnS NPs [11], the most common is the co-precipitation method [12–14], due to the advantage of simplicity, low-cost, hand-on and high-yielding production compared to the other methods [2, 11].

A common drawback that may arise when considering the implementation of ZnS in the form of very fine particles, required in some of the potential applications mentioned above, is the tendency of such particles to agglomerate due to the so-called Ostwald ripening [15]. A strategic way to overcome such an issue is the surface modification of inorganic nanomaterials through functionalization with organic capping agents, which demonstrates to be a very useful and hand-on tool to avoid particle agglomeration, thus improving the current properties and even to design new ones [16]. For instance, the synthesis of ZnS NPs using Schiff bases as capping agents has been intensively studied to improve the efficiency in photocatalytic applications [17]. The formation of ZnS NPs from chitosan laurate micellar solution that can be adopted as the reaction template [12], or the use of poly(N-vinyl-2-pyrrolidone to encapsulate ZnS NPs for tuning the optical properties [13] can also be mentioned as representative examples in the realm of surface organo-modification of ZnS NPs.

When introducing the capping agent in the targeted reaction process, in fact the stability of the oxidation states controlling the performance of Zn(II) ions in a large variety of useful catalytic applications is therefore attained [18]. Another important consequence induced through the arrest of particles agglomeration is the achievement of size selectivity during the co-precipitation reaction, by which the optical properties, such as the absorption, photoluminescence and band gap energy can be adjusted [19].

Aiming to investigate the yet less explored influence of other types of organic compounds as capping agents towards the surface functionalization of ZnS NPs with enhanced photoluminescent properties and band gap energy, in the present work we report the fabrication, through co-precipitation method, of a new series of ZnS NPs modified with tris(hydroxymethyl)aminomethane (TRIS) and 4-(2-hydroxyphenyl)-2-(morpholin-4-yl)-1,3-thiazole (DF) (Fig. 1). The two chosen organic compounds are regarded as structural units of importance in polyoxometalate chemistry, being used in biochemistry and molecular biology [20, 21]. Recognizing their coordinative aptitudes, the organic compounds TRIS and DF can thus be introduced into the surface modification practice of ZnS NPs in order to contribute both to reducing the particles agglomeration and size, as well as to attain enhanced photoluminescent properties.

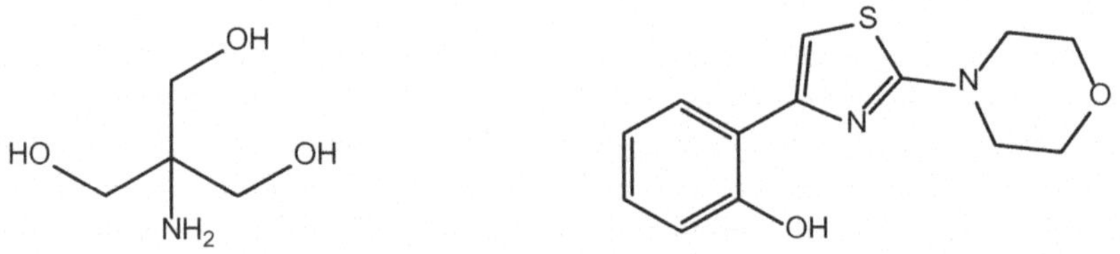

tris(hydroxymethyl)aminomethane (TRIS) 4-(2-hydroxyphenyl)-2-(morpholin-4-yl)-1,3-thiazole (DF)

Fig. 1. Chemical structure of the two capping agents used in the surface organo-modification of ZnS NPs.

2 Experimental

2.1 Materials

The organic compounds used as capping agents for the surface modification of ZnS NPs, tris(hydroxymethyl)aminomethane (TRIS) [21] and 4-(2-hydroxyphenyl)-2-(morpholin-4-yl)-1,3-thiazole (DF) [20], were synthesized following the methods reported in the literature. Zinc acetate dihydrate ($Zn(CH_3COO)_2{\cdot}2H_2O$, 98%), anhydrous sodium sulphide (Na_2S, 98%) and absolute ethanol (99.5%) and were purchased from Merck and used without further purification.

2.2 Synthesis of ZnS NPs

For the synthesis of unmodified ZnS NPs, 25 mL of aqueous solutions containing zinc acetate dihydrate (25 mM) and sodium sulphide (25 mM) were initially prepared and mixed (1:1, v/v) in a 100-mL round bottom flask, under continuous and vigorous magnetic stirring at 65 °C for 30 min. Unmodified ZnS NPs were formed as a white precipitate, which was cooled and collected by centrifugation (9000 rpm, 10 °C, 5 min), washed with two portions of ethanol, and finally dried at 100 °C in an oven for 3 h. For the synthesis of organo-modified ZnS NPs, 10 mL of ethanol solutions containing the capping agents TRIS and, respectively, DF (1 mM) were prepared and mixed with 25 mL of the aqueous solutions containing zinc acetate dihydrate (25 mM) under vigorous magnetic stirring at room temperature. After 30 min, 25 mL of the aqueous solution containing sodium sulphide (25 mM) were added drop by drop to the reaction mixtures containing the capping agents and zinc acetate, under vigorous magnetic stirring at room temperature for 30 min. The formed white (ZnS NPs modified with TRIS) and pale yellow (ZnS NPs modified with DF) precipitates were collected by centrifugation (9000 rpm, 10 °C, 5 min), washed with two portions of ethanol, and finally dried at 100 °C in an oven for 3 h. For the sake of simplicity, the obtained samples were labelled with ZnS for the unmodified nanoparticles, taken as the reference material, and ZnS-TRIS and ZnS-DF for the nanoparticles modified with TRIS and DF capping agents.

2.3 Characterization Methods

X-ray powder diffraction (XRPD) patterns were recorded on Bruker DRX-400 spectrometer (Germany). The XRPD measurements were taken at room temperature with a monochromatic radiation Cu $K_{\alpha 1}$ (λ = 1.54059 Å). The analyses were performed by using Bragg–Brentano geometry ("theta–2 theta") for angles between 10 and 80° 2θ degrees with a step size of 0.04° and acquisition time on step size of 3 s. Structural parameters of the Brown HT powder were evaluated by pattern matching, using the Full Prof software. The average crystallite size (L) of the obtained ZnS NPs was calculated following the Scherrer's equation:

$$B(2\theta) = (K \cdot \lambda) / (L \cdot \cos\theta) \tag{1}$$

where B is the full width at half maximum intensity (FWHM), λ is the X-ray wavelength, θ is the diffraction angle and K is the Scherrer's constant whose value for spherical particles is 0.89 [22].

The morphology and chemical composition of ZnS NPs were analyzed through scanning electron microscopy (SEM) with Philips scanning electron microscope XL 30 FEG Quanta 2000 operating at 15 kV, coupled with an energy dispersive X-ray spectroscopy (EDX) system. Specimens were prepared by dispersing the samples by sonication in ethanol and by depositing few drops of the suspensions on carbon-coated grids.

Fourier transform infrared spectroscopy (FTIR) analysis of ZnS NPs was performed with a Bruker Alpha spectrometer (Germany). The infrared spectra of samples were collected in the range 4000–500 cm^{-1}, with the resolution of 2 cm^{-1}.

Solutions of ZnS NPs (1 mM) were freshly prepared in absolute ethanol and analysed by optical absorption spectra using PG Instruments T70 UV-Vis spectrophotometer (Germany), in the 200–800 cm^{-1} wavelength range, using 1-mL quartz cells.

The pH and conductivity were evaluated at room temperature using a multi-parameter CONSORT C3021. The samples placed in hermetically sealed containers were initially analysed, and after 3 and 6 days of the reaction.

Solid state photoluminescence spectra were recorded with a Microplate reader with fluorescence spectrometer Infinite 200 PRO NanoQuant (Tecan, Männedorf, Switzerland), in the 400–800 nm range, upon excitation at 365 nm.

3 Results and Discussion

3.1 Synthesis and Physico-Chemical Analysis of ZnS NPs

The precursors zinc acetate dihydrate and sodium sulphide were dissociated into the corresponding cations (Zn^{2+}, Na^{+}) and, respectively, anions (CH_3COO^{-}, S^{2-}). Sodium and acetate ions remain in solution, while ZnS particles are formed due to the reaction between Zn^{2+} and S^{2-} ions, which subsequently can grow by consuming the smaller particles.

$$\mathrm{Zn}\left(\mathrm{CH_3COO}\right)_2 + \mathrm{Na_2S} \rightarrow 2\mathrm{CH_3COONa} + \mathrm{ZnS} \tag{2}$$

The growth of ZnS NPs can be stopped by the accumulation of sodium ions from Na_2S on the surface or by introducing into the reaction system a capping agent. Figure 2 presents the proposed mechanism of the surface capping of ZnS NPs with TRIS, which undergoes deprotonation at the hydroxyl groups in the basic ambient assured by the sodium ions, and subsequent covalent bonding with the eventual Zn^{2+} vacant sites from the surface of formed ZnS NPs. Figure 3, instead, presents the proposed mechanism of the surface capping of ZnS NPs with DF, using the oxygen donor atom from the morpholin moiety, whose position is more favorable to establish dative bonds with the Zn^{2+} vacant sites from the surface of formed ZnS NPs.

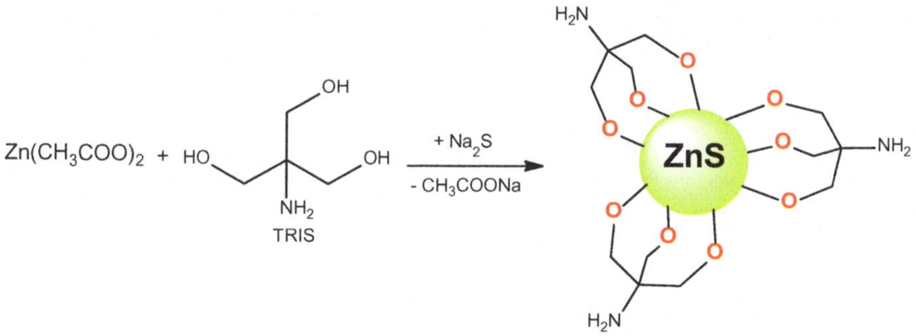

Fig. 2. Proposed mechanism for the formation of ZnS-TRIS NPs with core-shell fashion.

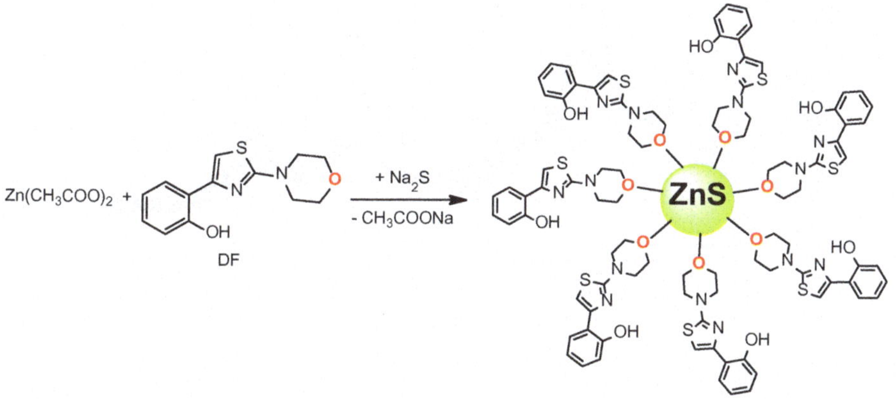

Fig. 3. Proposed mechanism for the formation of ZnS-DF NPs with core-shell fashion.

In both cases, core-shell structures are envisaged, with the terminal organic moieties from the used capping agents oriented outwardly so that to keep the ZnS NPs away from each other, thus preventing their agglomeration and further growth. The surface energy due to the nucleation of ZnS is very high and the surface is passivated by adsorption of anions from the solution [14, 23]. Therefore, the surface capping could be more favourable by oxygen atoms having electrons pairs. The accumulation of anions attracts cations towards the surface, thus suggesting a future application that is the immobilization of heavy metals at ZnS NPs surface [4].

Physic-chemical characterization of the obtained ZnS NPs was also performed by means of studies regarding the evolution in time of the pH and conductivity of solutions containing both unmodified and organo-modified ZnS NPs (Table 1). pH values slowly decreased by adding capping agents in ZnS NPs synthesis and an increasing of solutions' conductivity was observed. By the structural changes induced by the capping agents, the conductivity of ZnS NPs was improved. These modifications indicate an improvement of the electrical properties of ZnS NPs using the TRIS and DF capping agents in the synthesis process. In the freshly prepared solutions the observed increase

Table 1. Evolution of pH and conductivity of solutions containing unmodified and organo-modified ZnS NPs (1 mM).

	ZnS NPs		ZnS-TRIS		ZnS-DF	
Time (days)	pH	λ (μS/cm)	pH	λ (μS/cm)	pH	λ (μS/cm)
Fresh	6.22	1047	5.56	1156	5.91	1163
3	6.80	1229	4.61	1365	6.00	1325
6	5.73	1333	5.82	1385	6.23	1208

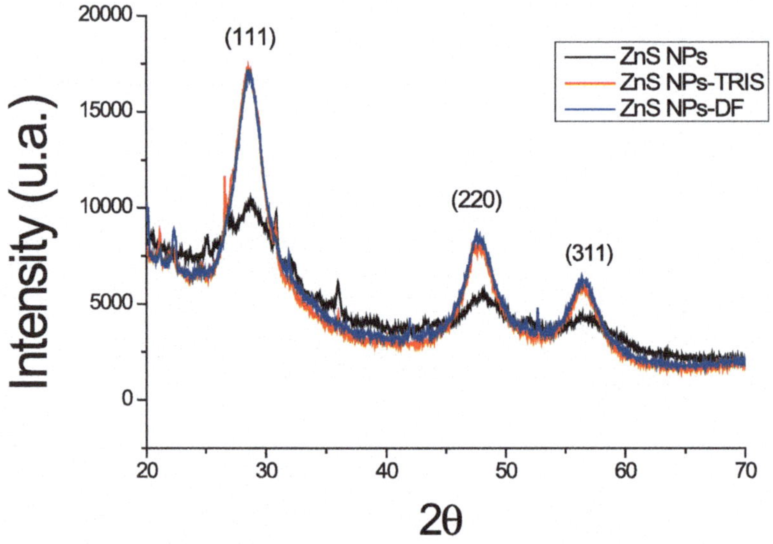

Fig. 4. XRPD patterns of ZnS, ZnS-TRIS and ZnS-DF nanoparticles.

in conductivity, from 1047 μS/cm (ZnS NPs) to 1156 μS/ (ZnS-TRIS) and 1163 μS/cm (ZnS-DF), can be due to the conducting nature of the charged nanoparticles dispersed in solution.

After 3 days an increase of the conductivity, from 1229 μS/cm for unmodified ZnS NPs to 1365 μS/cm for ZnS-TRIS and 1325 μS/cm for ZnS-DF, was observed. After 6 days of reaction almost the same values for both ZnS and ZnS-TRIS were obtained, which indicates a good efficiency of the transport of electrical charges in time, but also a good stability over time of the obtained samples.

3.2 Morpho-Structural Characterization of ZnS NPs

The XRPD patterns of unmodified and organo-modified ZnS NPs are presented in Fig. 4. The phase formation of unmodified ZnS NPs was confirmed by the peaks observed at 2θ values of 33°, 47.4°, 56.4° and 69.4°, which perfectly match the crystalline planes (111), (220), (311) and (400), respectively, corresponding to the face centred cubic structure of pure ZnS [2–5]. The same diffraction patterns were also obtained for ZnS NPs modified with TRIS and DF. The average crystallite size of the obtained ZnS NPs, calculated with the Scherrer's equation, underwent a decrease from 8.57 ± 1 nm in unmodified ZnS NPs, to 4.28 ± 1 nm in both organo-modified ZnS NPs. The reducing in the size of ZnS NPs (cca. 50%) using the TRIS and DF capping agents indicates an inhibition effect of the agglomeration during the formation of ZnS NPs [24].

Scanning electron microscopy with energy dispersive X-ray spectroscopy (SEM-EDX) was employed in view of getting useful information regarding the morphology, particle size and compositional analysis of the obtained ZnS samples. The micrograph of unmodified ZnS sample shows its consistency of particles with nearly spherical shape and size of *ca.* 40 nm, which in turn are organized into bigger agglomerates (Fig. 5A). The morphology of both organo-modified ZnS samples is similar to the unmodified ones, and their size is significantly decreased with respect to the unmodified particles, amounting to *ca.* 10 nm (Fig. 5B and C). This result allows to invoke the important effect of both capping agents in restricting the particles agglomeration, as well as in reducing the size of the obtained ZnS NPs.

EDX analysis confirmed the presence of the main elements in the unmodified ZnS NPs (Zn and S) and the appearance of carbon and nitrogen atoms in ZnS NPs modified with TRIS and DF capping agents (Fig. 6 and Table 2). The amount of oxygen present on the nanoparticles surface decreases when the capping agents are used for the analysis of nanoparticles. In addition, EDX analysis shows that the capping agents reduce the degree of oxidation of the nanoparticles, fact indicated by the decreases of the amount of oxygen. Another important observation is that the increase of the amount of carbon on the nanoparticles induces the decrease of the Zn and S atoms, thus indicating that the organic species are adsorbed on ZnS NPs surface.

FTIR spectroscopy was employed to assess both the successful production of ZnS NPs and the covalent attachment of the organic capping agents onto the ZnS surface (Fig. 7). The absorption observed at 590 cm^{-1} in both unmodified and organo-modified ZnS NPs is assigned to stretching vibration of Zn–S bonds of the cubic phase [25]. The broad band observed in the range 800–1000 cm^{-1} in unmodified ZnS NPs is assigned to the adsorption of carbon dioxide on the zinc sulphide surface, possibly coming from the atmosphere [26]. The peaks located at 1108 and 1248 cm^{-1} correspond to the stretching vibration of C–N and C–O bonds, which are common in both TRIS and DF capping agents present in the organo-modified ZnS NPs. Moreover, the peaks from the range

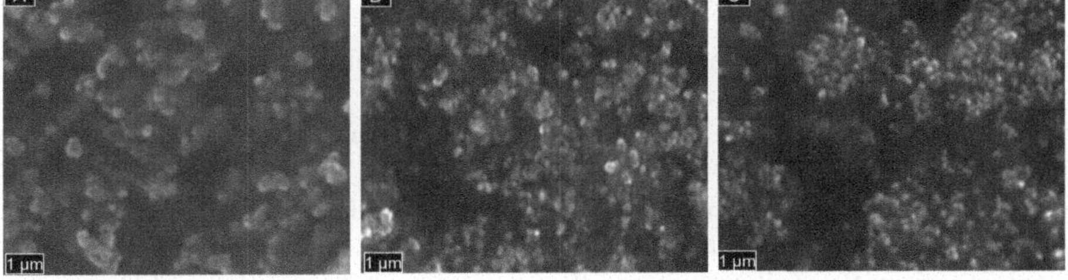

Fig. 5. SEM images of unmodified ZnS NPs (A), compared with TRIS- capped ZnS NPs (B) and DF-capped ZnS NPs (C).

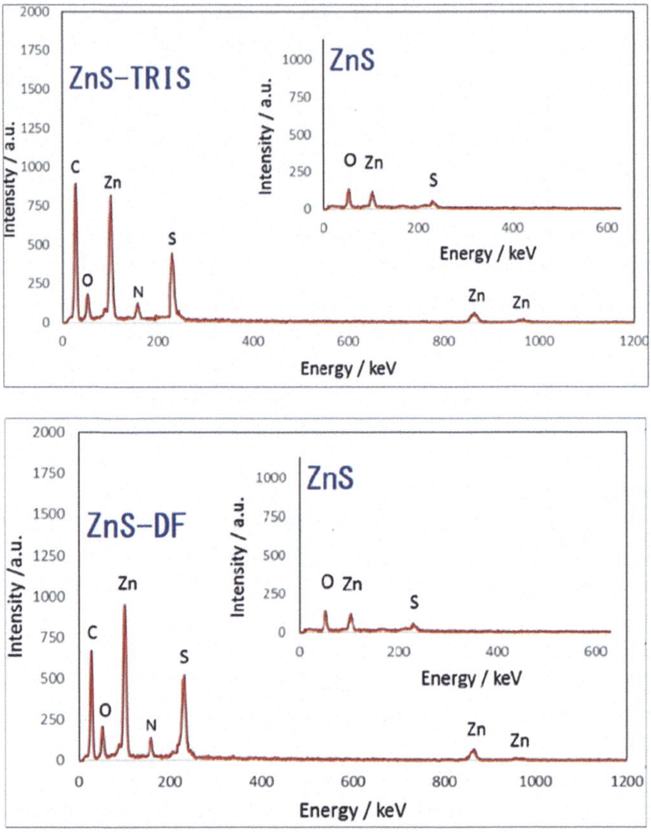

Fig. 6. EDX analysis of ZnS NPs modified with TRIS (top) and DF (bottom). Insert: EDX analysis of unmodified ZnS NPs.

Table 2. Composition of ZnS NPs samples from EDX analysis.

Sample	Zn (wt%)	S (wt %)	O (wt %)	C (wt %)	N (wt %)
ZnS	55.60	26.30	18.10	–	–
ZnS-TRIS	23.88	7.93	6.12	59.03	3.04
ZnS-DF	25.72	10.49	7.55	51.01	5.27

1242–1552 cm^{-1} may cover both the asymmetric and symmetric stretching of the carboxylate group coming from the acetate anions, not fully removed even after insistent washings [27], along with the vibration of C–S bond from the structure of DF capping agent [28].

The peaks observed in the range 800–1000 cm^{-1} in both TRIS- and DF-capped ZnS NPs are attributed to the vibrational modes of Zn–O bonds established between the Zn^{2+} vacant sites from the ZnS surface and the oxygen atoms from TRIS and DF capping agents during the formation of organo-modified ZnS NPs in the core-shell fashion [19].

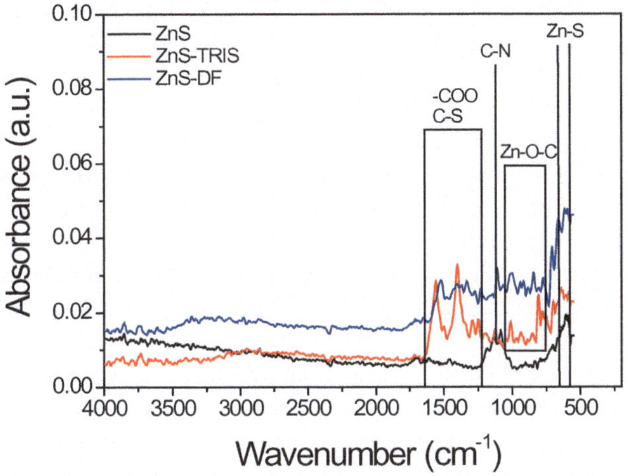

Fig. 7. FTIR spectra of ZnS NPs and organo-modified ZnS NPs.

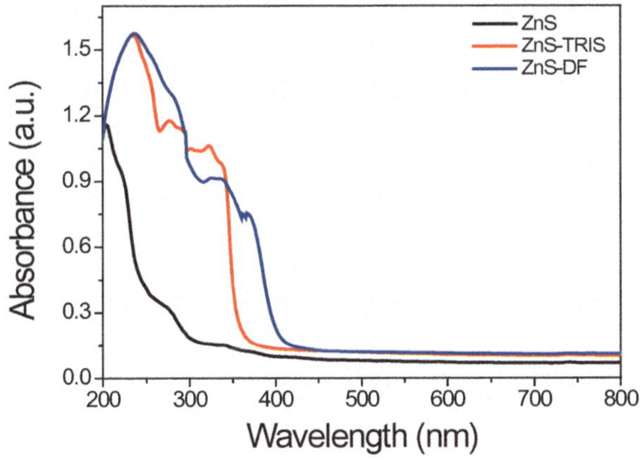

Fig. 8. UV-Vis absorption spectra of ethanol solutions (1 mM) containing unmodified and organo-modified ZnS NPs.

3.3 Optical Properties

The ethanol solutions containing the as-prepared unmodified and organo-modified ZnS NPs were analyzed by UV-vis absorption spectroscopy, in order to observe the eventual changes and the effect of capping agents upon their formation, as well as upon the band gap energy (Fig. 8). The absorption edge for unmodified ZnS NPs was observed at 334 nm, thus confirming their successful formation in the adopted reaction conditions [29, 30]. In both organo-modified ZnS NPs the absorption edge underwent a non-negligible blue shift to 324 nm, as a consequence of the quantum confinement effects involving the photogenerated electron–hole carriers, induced by the nanoparticles size decrease. Such a blue shift could be attributed to transitions that occur from the valence band to the conduction band [31, 32]. Several absorption bands, also observed for both organo-modified ZnS NPs in the range 230–350 nm, may be due to the electronic transition occurring at the level of the organic components used in the surface functionalization of zinc sulphide.

The optical band gap (E_g) of the obtained ZnS NPs was calculated with the formula $E_g = 1240/\lambda$ [33], where λ is the wavelength of the absorption edge observed in the UV-vis spectra. As such, while the band gap energy of unmodified ZnS NPs is equal to 3.71 eV, for both TRIS- and DF-capped ZnS NPs the value increased to 3.83 eV. These results confirm the general trend observed in the literature, according to which the band gap energy increases when the nanoparticles size decreases, as a result of quantum size effects, as was also observed in, *e.g.*, mercaptoethenol-capped ZnS NPs [29, 31] and thiophenol-capped ZnS quantum dots [34]. Such a band gap increase can be due to a certain shrinking of the valence and conduction bands determined by the electron injection coming from the lone pairs belonging to the oxygen atoms from the two organic capping agents bound to the ZnS surface.

Upon exposure to UV light, at the excitation wavelength of 365 nm, both unmodified and organo-modified ZnS NPs display photoluminescent properties with different emission intensities, and their photoluminescent spectra, measured at room temperature in the solid state, are collectively gathered in Fig. 9.

All ZnS samples showed broad yellow-to-green emission bands, whose maximum vary from 580 nm in unmodified ZnS to 560 nm in TRIS- and DF-modified ZnS NPs. The observed blue shift phenomenon, running from unmodified to organo-modified ZnS NPs, is also accompanied by a progressive enhancement in the photoemission intensity, as a consequence of the particle size decrease induced by the surface capping. To the best of our knowledge, such yellow-to-green emissions are relatively rare for semiconductor ZnS NPs, and may originate from the recombination phenomenon of the electrons and holes trapped at the surface state defects, as were also observed in manganese-doped ZnS NPs [32] or in EDTA-capped ZnS NPs [35].

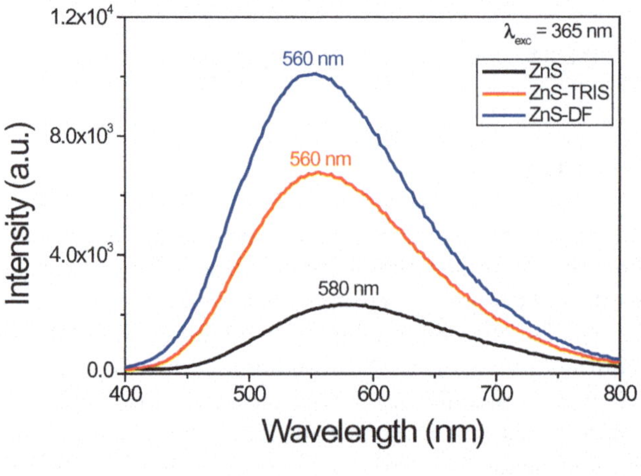

Fig. 9. Photoluminescent spectra of both unmodified and organo-modified ZnS NPs.

Surface modification of ZnS NPs with the organic capping agents like TRIS and DF can thus be regarded as a remarkable strategy which leads to a significant increase in the number of effective surface luminescence centers, attainable by the reduction of particle agglomeration or aggregation.

Consequently, this leads to the enhancement of photoemission intensity in the obtained organo-modified ZnS NPs. The eventual influence of the nature of organo-modifiers upon the photoemission intensity of surface-capped ZnS NPs should also be taken into consideration. In fact, the ZnS sample capped with DF, possessing chromophore groups by means of C=C and C=N double bonds in the aromatic moieties, induced the most intense emission, compared to the ZnS sample capped with TRIS.

4 Conclusions

In the present work, a simple, low-cost and high-yielding chemical co-precipitation method for the fabrication of very fine zinc sulphide nanoparticles was reported, by introducing into the surface modification practice the organic capping agents tris(hydroxymethyl)aminomethane (TRIS) and 4-(2-hydroxyphenyl)-2-(morpholin-4-yl)-1,3-thiazole (DF). The change in the morphological and optical properties of core–shell nanoparticles could be observed by varying the nature of capping agents. The role of both capping agents was to stabilize and prevent the obtained nanoparticles from agglomeration, thus leading to a significant influence regarding the improvement of their structural and optical properties. The morpho-structural characterization techniques used in the present work allowed to confirm the important effect of the chosen capping agents in the preparation of ZnS NPs with size reduction from *ca.* 40 nm in unmodified ZnS NPs to *ca.* 10 nm in ZnS NPs modified with either TRIS or DF capping agents. Such an effect came along with enhanced yellow-to-green photoluminescence of the organo-modified ZnS NPs and an increase of the band gap energy from 3.71 eV in unmodified ZnS to 3.83 eV in organo-modified ZnS NPs. Based on these results, the adopted synthetic method can be exploited for adjusting the particle size and photoluminescent properties of ZnS NPs for their implementation into new generations of optoelectronic devices, such as hybrid light-emitting diodes.

References

1. Kaur, N., Kaur, S., Singh, J., Rawat, M.: A review on zinc sulphide nanoparticles: from synthesis, properties to applications. J. Biomed. Nanotechnol. **1**(1), 1–5 (2016)
2. Fang, X., Zhai, T., Gautam, U.K.: ZnS nanostructures: from synthesis to applications. Prog. Mater. Sci. **56**(2), 175–287 (2011)
3. Porta, F.A., Nogueira, A.E., Gracia, L., Pereira, W.S., Botelho, G., Mulinari, T.A., Andrés, J., Longo, E.: An experimental and theoretical investigation on the optical and photocatalytic properties of ZnS nanoparticles. J. Phys. Chem. Solids. **103**, 179–189 (2017)

4. Ayodhya, D., Veerabhadram, G.: Fabrication of Schiff base coordinated ZnS nanoparticles for enhanced photocatalytic degradation of chlorpyrifos pesticide and detection of heavy metal ions. J. Mater. **5**(3), 446–454 (2019)
5. Viswanath, R., Bhojya Naik, H.S., Arun Kumar, G., Suresh Gowda, I.K., Yallappa, S.: Tuneable luminescence properties of EDTA-assisted ZnS:Mn nanocrystals from a yellow-orange to a red emission band. Luminescence. **32**(7), 1212–1220 (2017)
6. Zhong, F., Wu, Z., Guo, J., Jia, D.: Ni-doped ZnS nanospheres decorated with au nanoparticles for highly improved gas sensor performance. Sensors. **18**(9), 2882 (2018)
7. Mintcheva, N., Gicheva, G., Panayotova, M., Kulinich, S.A.: Room-temperature synthesis of ZnS nanoparticles using zinc xanthates as molecular precursors. Materials. **13**, 171 (2020)
8. Pratap, S., Prasad, J., Kumar, R., Murari, K., Singh, S.S.: Preparation and characteristics of IR window grade zinc sulphide powder. Def. Sci. J. **46**(4), 215–221 (1996)
9. Yang, X., Feng, G., Yang, C., Wang, S., Zhou, S.: Random lasing in Cr:ZnS microstructure-textured grooves. Laser Phys. **26**(4), 045702 (2016)
10. Chien, H.C., Cheng, C.Y., Mao, M.H.: Continuous wave operation of SiO_2 sandwiched colloidal CdSe/ZnS quantum dot microdisk lasers. IEEE J. Sel. Top. Quant. Electron. **23**(5), 1–5 (2017)
11. Dewi, R.P., Nurdiana, A., Astuti, L., Ragadita, R., Nandiyanto, A.B.D.: Review: synthesis of zinc sulfide nanoparticles by various methods. Arab. J. Chem. Environ. Res. **08**(02), 355–371 (2021)
12. Khiew, P.S., Radiman, S., Huang, N.M., Soot Ahmad, M.D., Nadarajah, K.: Preparation and characterization of ZnS nanoparticles synthesized from chitosan laurate micellar solution. Mater. Lett. **59**(8–9), 989–993 (2005)
13. Ghosh, G., Naskar, M.K., Patra, A., Chatterjee, M.: Synthesis and characterization of PVP-encapsulated ZnS nanoparticles. Opt. Mater. **28**, 1047–1053 (2006)
14. Alivisatos, A.P.: Semiconductor clusters, nanocrystals, and quantum dots. Science. **271**, 933–937 (1996)
15. Ratke, L., Voorhees, P.W.: Growth and Coarsening: Ostwald Ripening in Material Processing. Springer, New York (2002)
16. Wieszczycka, K., Staszak, K., Woźniak-Budych, M.J., Litowczenko, J., Maciejewska, B.M., Jurga, S.: Surface functionalization—the way for advanced applications of smart materials. Coord. Chem. Rev. **436**, 213846 (2021)
17. Pourjafari, D., Saeednia, S., Iranmanesh, P., Hatefi Ardakani, M.: Preparation and photocatalyst properties of zinc sulfide using a distinct sulfur source: 1-Benzylidenethiourea. J. Clust. Sci. **30**, 571–580 (2019)
18. Dalia, S.A., Afsan, F., Hossain, M.S., Khan, M.N., Zakaria, C., Zahan, M.E., Ali, M.: A short review on chemistry of Schiff base metal complexes and their catalytic application. Int. J. Chem. Stud. **6**(3), 2859–2867 (2018)
19. Sharma, M., Kumar, S., Pandey, O.P.: Study of energy transfer from capping agents to intrinsic vacancies/defects in passivated ZnS nanoparticles. J. Nanopart. Res. **12**, 2655–2666 (2010)
20. Chirita, P., Duinea, I.M., Sandu, A., Birsa, L., Sarbu, L., Baibarc, M., Sava, F., Popescu, M., Matei, E.: Inhibitory effect of three phenacyl derivatives on the oxidation of sphalerite (ZnS) in air-equilibrated acidic solution. Corros. Sci. **138**, 154–162 (2018)
21. Duinea, I.M., Carac, G., Dabuleanu, I.D., Petcu, M.A., Sarbu, L.G., Birsa, L., Chirita, P.: Troilite (FeS) oxidation in the presence of a newly synthesized TRIS-based base. Revista de Chimie (Bucharest). **70**(7), 2639–2642 (2019)
22. Klug, H.P., Alexander, L.E.: X-Ray Diffraction Procedures for Polycrystalline and Amorphous Materials. Wiley, New York (1974)
23. Hudlikar, M., Joglekar, S., Dhaygude, M., Kodam, K.: Latex-mediated synthesis of ZnS nanoparticles: green synthesis approach. J. Nanopart. Res. **14**, 865 (2012)

24. Kumara, C., Armstrong, B., Lyo, I., Leeb, H.W., Qu, J.: Organic-modified ZnS nanoparticles as a highperformance lubricant additive. RSC Adv. **13**, 7009–7019 (2023)
25. Murugadoss, G., Rajesh Kumar, M.: Synthesis and optical properties of monodispersed Ni^{2+}-doped ZnS nanoparticles. Appl. Nanosci. **4**, 67–75 (2014)
26. Coenen, K., Gallucci, F., Mezari, B., Hensen, E., van Sint Annaland, M.: An in-situ IR study on the adsorption of CO_2 and H_2O on hydrotalcites. J. CO2 Util. **24**, 228–239 (2018)
27. Nakamoto, K.: Infrared and Raman Spectra of Inorganic and Coordination Compounds, 6th edn. Wiley, Hoboken, NJ (2009)
28. Kuppayee, M., Vanathi, G., Nachiyar, K., Ramasamy, V.: Synthesis and characterization of Cu^{2+} doped ZnS nanoparticles using TOPO and SHMP as capping agents. Appl. Surf. Sci. **257**, 6779–6786 (2011)
29. Kumbhojkar, N., Nikesh, V.V., Kshirsagar, A., Mahamuni, S.: Photophysical properties of ZnS nanoclusters. J. Appl. Phys. **88**, 6260–6264 (2000)
30. Raevskaya, A.E., Korzhak, A.V., Stroyuk, A.L., Kuchmii, S.Y.: Spectro-optical and photochemical properties of ZnS nanoparticles. Theor. Exp. Chem. **41**(2), 111–116 (2005)
31. Tamrakar, R., Ramrakhiani, M., Chandra, B.P.: Effect of capping agent concentration on photophysical properties of zinc sulfide nanocrystals. Open Nanosci. J. **2**, 12–16 (2008)
32. Murugadoss, G., Ramasamy, V.: Synthesis, effect of capping agents and optical properties of manganese-doped zinc sulphide nanoparticles. Luminescence. **28**(1), 69–75 (2013)
33. Dharma, J., Pisal, A.: Simple method of measuring the band gap energy value of TiO_2 in the powder using UV/vis/NIR spectrometer, pp. 1–4. Perkin Elmer, Inc., Shelton (2009)
34. Mahamuni, S., Khosravi, A.A., Kundu, M., Kshirsagar, A., Bedekar, A., Avasare, D.B., Singh, P., Kulkarni, S.K.: Thiophenol-capped ZnS quantum dots. J. Appl. Phys. **73**(10), 5237–5240 (1993)
35. Osman, M.A., Othman, A.A., El-Said, W.A., Abd-Elrahim, A.G., Abu-Sehly, A.A.: Thermal annealing and UV irradiation effects on structure, morphology, photoluminescence and optical absorption spectra of EDTA-capped ZnS nanoparticles. J. Phys. D. Appl. Phys. **49**(5), 055304 (2015)

Study on Water Quality at Sungai Batu Feringghi, Pulau Pinang by Using Qual2K Modelling and GIS

Ong Jing Hong[1], Mohamad Anuar Kamaruddin[1] (✉), Rasyidah Alrozi[2], Norazian Mohamed Noor[3], and Mohd Remy Rozainy Mohd Arif Zainol[4]

[1] Environmental Technology Division, School of Industrial Technology, Universiti Sains Malaysia, Gelugor, Pulau Pinang, Malaysia
anuarkamaruddin@usm.my
[2] College of Engineering, UiTM Cawangan Pulau Pinang, Permatang Pauh, Pulau Pinang, Malaysia
[3] Faculty of Civil Engineering & Technology, Universiti Malaysia Perlis (UniMAP), Perlis, Malaysia
[4] River Engineering and Urban Drainage Research Centre (REDAC), Universiti Sains Malaysia, Gelugor, Pulau Pinang, Malaysia

Abstract. Downstream of Sungai Batu Ferringhi is connected to the popular tourism spot of Batu Ferringhi Beach in Malaysia. The water quality deterioration in Sungai Batu Ferringhi directly affects the seawater quality, subsequently impacting fisheries and tourism activities. This study aims to identify both point and diffuse pollution sources in Sungai Batu Ferringhi and simulate water quality during normal flow and low flow events using QUAL2K. Concentrations of biochemical oxygen demand (BOD), ammoniacal nitrogen (NH_3-N), total suspended solids (TSS), and total phosphorus are the focal water quality parameters. 6 water quality stations and 2 point sources were identified, and samples were collected from each. As per the National Water Quality Standards (NWQS), Sungai Batu Ferringhi falls under Class III. The baseline model was initially simulated and calibrated using a trial-and-error method. A 7Q10 low flow event was simulated by modifying water flow. During both normal flow and 7Q10 low flow events, most upper and middle stream sections of Sungai Batu Ferringhi met Class III standards, except for total phosphorus concentration. However, certain water quality parameters like BOD and total phosphorus failed to meet Class III standards, particularly in the middle and lower stream sections during both flow scenarios. A noticeable increase in BOD and total phosphate concentrations in the middle stream during low flow events impacted the river's water quality. Both point sources had negligible impacts on Sungai Batu Ferringhi's water quality. Diffuse sources, potentially from forest runoff, urban runoff, and agricultural runoff, are identified as the primary contributors of pollutants to Sungai Batu Ferringhi.

Keywords: Bioremediation · Refinery oily sludge · Food waste digestate · Hydrocarbon · Seed germination

A. V. Sandu et al. (Eds.): ICIR EUROINVENT 2024, SPM 64, pp. 39–52, 2025.
https://doi.org/10.1007/978-3-031-81198-2_4

1 Introduction

1.1 Research Background

Rivers are always the main freshwater sources for human consumption and many activities including irrigation, sanitation and industrial use [1]. For most conditions, the water in the river upstream is not polluted. However, as the water flows downstream, the river will receive many point and non-point contamination sources which eventually lead to the major degradation of river water quality [2]. Even though human depends heavily on the freshwater supply from the rivers, many anthropogenic activities have contributed to the release of a significant amount of pollutants into the river which resulted in serious contamination of the surface water [3]. Surface water is highly vulnerable to pollution as it will affect not only humans but also many living organisms, especially aquatic organisms. The degradation of surface water resources is mainly caused by urbanization, the rapid development of industrial activities, uncontrolled use of chemical fertilizers, intensive farming, and improper discharge of untreated effluents and sewage [4]. It is critical to always monitor the river water quality closely before taking the water to serve any purpose.

1.2 Problem Statement

In Malaysia, the water quality index (WQI) is widely utilized to quantify and represent the overall water quality status. The application of WQI can assist decision- makers in determining the suitability of a water body for various purposes. Sungai Batu Ferringhi has been reported to be a clean river from the year 2017 till the year 2021 with both river water quality monitoring stations categorized as mainly class II and class I based on the WQI value. However, many concerns still arise because the WQI value does not reflect the actual conditions of every water quality parameter [5]. In Malaysia, the WQI calculation does not take into account many parameters such as the presence of faecal coliform and *Escherichia coli* (E.coli) in the water and the aforementioned parameters have been found in abundant amounts, especially in sewage effluent. It is crucial to determine every important water quality parameter before we deduce that the river water is safe for any purpose instead of only referring to the WQI value. A case study related to the analysis of water quality in Sungai Batu Ferringhi conducted by Rajesh Nithyanandam in the year 2014 [6] indicated a 900 cfu/100 mL concentration of E.coli in the river water even though the WQI value is under Class II as per NWQS. The river water still needs to be treated using the conventional method before any recreational purpose which only involves body contact.

The determination of pollution sources around the river is also necessary to maintain the river water quality at a safe level. Sungai Batu Ferringhi used to receive the effluent released from a nearby sewage treatment plant run by IWK. However, ever since the Batu Ferringhi Beach pollution in 2014, the management of the sewage treatment plant has decided to build another piping system to direct the final effluent into the sea without discharging into Sungai Batu Ferringhi. The remaining point pollution sources of Sungai Batu Ferringhi are the effluents discharged from the hawker stalls, residential areas and hotels nearby. The non-point pollution sources could be the urban runoff and runoff within the undeveloped forest area near the upstream of the river. These point and non-point pollution sources may pose a different level of threat towards the water quality of Sungai Batu Ferringhi depending on the characteristics and volume of the discharged effluents.

It is necessary to conduct a water quality modelling on Sungai Batu Ferringhi to ensure that the point and non-point pollution sources will not cause a major degradation of the river water quality. Sungai Batu Ferringhi must not be polluted since it is connected to Batu Ferringhi Beach which will directly influence the nearby tourism activities and possibly affect the marine ecosystem which may cause a decline in fisherman's income. Therefore, the QUAL2K model has been selected to study the impact of the identified point and non-point pollution sources towards Sungai Batu Ferringhi for better water management.

1.3 Research Objectives

i. To identify the point and non-point pollution sources of Sungai Batu Ferringhi by using hydraulics and hydrology field measurement and secondary data.
ii. To simulate the changes in the Sungai Batu Ferringhi water environment due to the point and non-point pollution sources based on different water quality parameters under normal flow and low flow (7Q10).

2 Literature Review

2.1 River Water Quality Status

The Department of Environment (DOE) in Malaysia conducts river water quality monitoring, employing both manual and continuous monitoring methods. In the process of manual monitoring, designated stations were used for the collection of water samples. Subsequent to sample collection, in-situ measurements and laboratory analyses were conducted to assess the physical, chemical, and biological attributes of the water samples. The Water Quality Index (WQI) is computed by taking into account several water quality parameters such as biochemical oxygen demand (BOD), chemical oxygen demand (COD), ammoniacal nitrogen, suspended solids, dissolved oxygen and pH. WQI is commonly employed to assess the extent of pollution and the associated suitability for various water uses as outlined in the National Water Quality Standards (NWQS) [7]. In 2021, 73% of monitored rivers exhibited a clean WQI value, while 24% were categorized as slightly polluted, and 3% were deemed polluted. In Penang, Sg. Chempedak is recorded as the most polluted river since it has the lowest WQI value which falls under class IV as per NWQS. As mentioned by DOE (2022), BOD, ammoniacal nitrogen, and suspended solids are the primary water quality parameters of interest because they are closely linked to pollution resulting from the release of pollutant loads, either from point sources or non-point sources, which can impact the quality of river water.

2.2 Water Pollution Load Analysis

When Penang has an estimated BOD load of 62.32 tonnes per day, ranking it as the fourth highest among all states in Malaysia. The estimated ammoniacal nitro-gen load in Penang is recorded at 16.36 tonnes per day, making it the fifth highest among other states in Malaysia. Additionally, the estimated load of suspended solids in Penang is 120.35 tonnes per day. Dilution serves as the principal mechanism through which the pollutant concentrations released from various point and non-point sources are decreased [8]. To evaluate the impact of a low-flow event on a river stream, the dilution factor can be computed by comparing the flow of the stream that receives pollutants to the flow of the discharged pollutants [9]. A high dilution factor implies that the river stream will experience a lower concentration of pollutants. Conversely, when the dilution factor is low, the river stream will exhibit a higher concentration of pollutants. A higher concentration of pollutants may disrupt the aquatic ecosystem and affect the survival of aquatic organisms. A low flow event is commonly considered in water quality management applications such as allocating the waste loads and discharge permits to a stream. Examining low-flow conditions enables water supply planning to determine permissible water transfers and withdrawals from a stream without giving adverse impacts towards the aquatic environment. Usually, water quality models utilize low-flow analysis as a standard design flow to simulate a worst-case scenario (IPASA, 2008).

3 Material and Method

3.1 River Water Sample Collection

Water quality sampling was necessary to obtain baseline values for various water quality parameters in Sungai Batu Ferringhi. Samples were collected from six (6) distinct water quality stations along Sungai Batu Ferringhi on September 16, 2023 (Table 1).

Triplicate samples are collected from each water quality station using the grab sampling method, which includes both the hand-dipped and pail sampling techniques.

Table 1. Coordinates of water quality stations.

Point	Coordinates	
	Latitude	Longitude
W1	5.461095	100.2423
W2	5.462173	100.2429
W3	5.464287	100.2429
W4	5.465832	100.2429
W5	5.466658	100.242
W6	5.467439	100.2405
P1	5.466525	100.2421
P2	5.466969	100.2413

On the sampling day, a total of two (2) point sources discharging water or wastewater into Sungai Batu Ferringhi are identified. Duplicate samples are collected from each identified point source using the same grab sampling method. To capture the normal flow of the river more accurately, sampling was carried out only after conducting multiple site visits to assess the water flow rate. The samples were collected during favorable weather conditions, with no significant wind presence.

To ensure an unbiased representation of baseline water quality, sampling was refrained from during rainfall and within the initial 24 hours after rain. During precipitation, pollutants from the surroundings may be transported into the river through runoff.

In the presence of strong winds, certain constituents might be displaced towards one side of the waterbody or undergo mixing, potentially aiding in their quicker dissipation [10].

3.2 Laboratory Analysis

All water quality laboratory analyses were conducted in Lab 262, located in the School of Industrial Technology, USM. The determination of 5-day biochemical oxygen demand (BOD5) could not be conducted due to the low accuracy of the dissolved oxygen meter. Half of each water sample was delivered to Bio Synergy Laboratories Sdn Bhd. for the determination of BOD5 concentration. For deter-mining the concentration of total suspended solids (TSS), the standard method outlined in APHA Standard Method for Examination of Water and Wastewater Section 2540 D titled "Total Suspended Solids Dried at 103–105 °C" was followed. For determining the concentration of COD, NH3-N and total phosphorus, DR900 which is a multiparameter colorimeter manufactured by HACH was used by referring to the manual provided by the manufacturer. The average value of the results was calculated for each station and point source to represent the water quality data at those locations. Characteristics physio-chemical of OS and FWD.

3.3 Modelling Tools

The modelling software that have been chosen in this study is QUAL2K. QUAL2K offers a comprehensive approach to analyse and simulate water quality dynamics in one-dimensional within rivers and streams [11]. This model presupposes specific elements, including:

- In a one-dimensional water flow setting, the channel exhibits thorough mixing both vertically and laterally.
- For hydraulics, a steady state is maintained, and the simulation incorporates non-uniform and consistent flow.
- The diel heat budget aspect involves simulating the heat budget and temperature in response to meteorological conditions over a diel time scale.
- Diel water-quality kinetics pertain to the simulation of all water quality variables within a diel time scale.
- A crucial assumption is made, which entails considering that the base-line data collected accurately represents normal flow conditions.

3.4 Segmentation

Before data input, the study area of Sungai Batu Ferringhi needs to be divided into several reaches with similar hydraulic characteristics. Subsequently, each reach is further segmented into a series of equally-spaced elements. In QUAL2K, an element represents the model's fundamental computational unit, comprising an equal length subdivision of a reach [11]. In this study, the number of elements in each reach is specified to be 5. Table 2 shows the reach data of Sungai Batu Ferringhi whereas Fig. 1 illustrates the schematic diagram of Sungai Batu Ferringhi. The upstream border of Sungai Batu Ferringhi is 0.991 km from the downstream.

3.5 Data Input

All the measured data, including the hydraulic and water quality data, are insert-ed into the QUAL2K model accordingly. To determine the depth and velocity of an element, the Manning Equation is used in this study. The equation is readily incorporated into the model with the variables to be input. Each element within a specific reach can be conceptualized as a trapezoidal channel as shown in Fig. 2. By assuming a steady state flow condition, the relationship between the flow and depth [11].

Table 2. Segmentation details.

Reach Number	Headwater Reach	Reach Length (km)	Location		Element Number > = 1
			Upstream (km)	Downstream (km)	
1	Yes	0.14	0.991	0.851	5
2		0.26	0.851	0.595	5
3		0.22	0.595	0.371	5
4		0.19	0.371	0.184	5
5		0.18	0.184	0.000	5

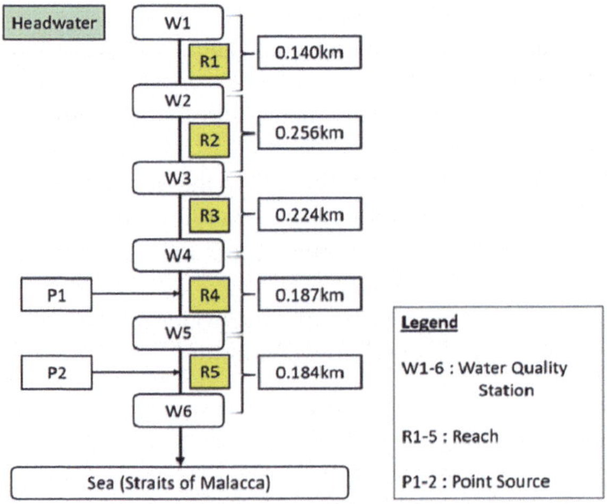

Fig. 1. Headwork, reach, and point source information.

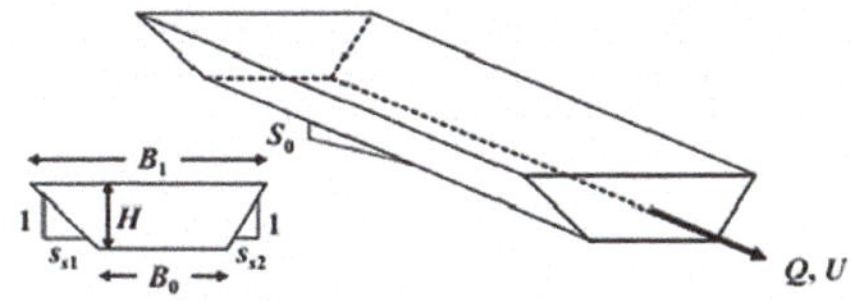

Fig. 2. Modelling framework.

Fig. 3. L-moment region in Peninsular Malaysia.

3.6 7Q10 Low Flow Analysis

The 7Q10 low flow event is commonly employed to predict a river's worst-case scenario, considering the reduction in volumetric flow rate. As there is a lack of long- term flow rate data for Sungai Batu Ferringhi, the 7Q10 flow rate is estimated through a set of procedures outlined in the Hydrological Procedure (HP) No.12, published by the Malaysia Department of Irrigation and Drainage (DID) and revised in 2015. This method for estimating low flow of a selected duration and frequency is applicable to ungauged rivers in Peninsular Malaysia. The catchment area of the study site is determined using GIS software, including Google Earth Pro. Annual mean rainfall data for the Batu Ferringhi area is collected from a weather forecast website for the period from January 2023 to December 2023. The L-moment region in Peninsular Malaysia is depicted in Fig. 3, extracted from HP No.12. The study area is within Region 9. The growth factors for Region 9 are given in Table 3.

Table 3. Growth factors for Region 9.

Flow duration days	Return period, years				
	2	5	10	20	50
1	0.924	0.525	0.280	0.121	0.011
4	0.930	0.555	0.328	0.170	0.056
7	0.934	0.589	0.345	0.177	0.258
30	0.947	0.605	0.359	0.187	0.062

Table 4. Geometric and hydraulic data.

Station	Width (m)	Depth (m)	Velocity (m/s)	Flow rate (m^3/s)
W1	2.0	0.3	0.125	0.0750
W2	2.0	0.3	0.131	0.0800
W3	1.7	0.7	0.093	0.1100
W4	6.4	0.8	0.076	0.3891
W5	11.1	3.6	0.048	1.9110
W6	12.7	3.7	0.050	2.3500
P1	–	–	–	0.0001
P2	–	–	–	0.0003

4 Result and Discussion

4.1 Geometric and Hydraulic Data

Table 4 lists out the geometric and hydraulic data measured in-situ during the sampling trip. The water flow rate at each station can be computed by multiplying the width of the river section by the water depth and then by the velocity of the river water. The river width undergoes a significant increase as it progresses downstream, from an initial width range of 1.7 m–2.0 m to a subsequent range of 6.4 m–12.7 m, particularly at points W4, W5 and W6, which are in close proximity to the outfall. Significant increments in water depth are noticeable at points W5 and W6, transitioning from an initial depth range of 0.3 m–0.8 m to a subsequent range of 3.6 m–3.7 m. The water velocity maintains a relatively consistent pattern across all stations, showing limited fluctuations within a range of 0.048 m/s to a maximum of 0.131 m/s. The water flow rate is observed to increase from the upstream to the downstream sections of the river. The flow rate varies between 0.0750 m^3/s and 2.3500 m^3/s. Elevated flow rates are evident at stations, notably W5 with a flow rate of 1.9110 m^3/s and W6 with a flow rate of 2.3500 m^3/s, situated close to the downstream section of the river.

The volumetric discharge rates for each point source are determined by dividing the volume of the collected effluent by the time taken for collection. Both point sources release a relatively small volume of effluent into the river, ranging from 0.0001 m^3/s to 0.0003 m^3/s. The effluent from P1 was collected from a pipe with an unidentified source. Potential sources of the effluent include nearby residential areas and restaurants, making them primary suspects. The effluent from P2 was also collected from a

Table 5. Water quality results for each station and point source.

Parameter	Water Quality Station					Point Source		
	W1	W2	W3	W4	W5	W6	P1	P2
T (°C)	26.91	27.01	27.34	27.76	28.71	27.30	27.28	28.99
pH	7.25	6.71	6.56	6.48	6.36	7.37	7.72	6.00
DO (mg/L)	4.25	4.88	4.60	4.63	4.45	4.30	3.86	6.47
BOD_5 (mg/L)	6	6	11	20	24	25	12	744
COD (mg/L)	18	18	17	16	29	25	42	647
Total Phosphorus (mg/L)	0.30	0.35	0.50	0.90	0.95	1.09	0.82	6.10
NH_3-N (mg/L)	0.01	0.01	0.02	0.03	0.03	0.03	0.27	0.90
TSS (mg/L)	17	14	23	42	45	41	4	246

pipe, located adjacent to the Hard Rock Hotel. Potential sources of this effluent include nearby hotel industries, particularly the Hard Rock Hotel itself. The effluent might be discharged from the hotel kitchen, potentially associated with the Hard Rock Café or from the water facilities such as swimming pool, as inferred from a logical deduction of the hotel layout.

4.2 Water Quality Data

Table 5 provides a summary of the water quality results for the samples collected from each station and point source. Average values for each parameter have been calculated. Using these average results for each parameter, the water quality station will be classified according to the standards outlined in the National Water Quality Standards (NWQS) for Malaysia. Table 5 displays the classification results of each parameter for each water quality station and point source, as per the NWQS. According to the water classes and designated uses specified in the NWQS, Sungai Batu Ferringhi is categorized as Class III. This classification is assigned because the river water is utilized for water supply and fishery purposes. Water Supply Class III implies that the river water requires significant treatment before it can be used for most purposes. Fishery Class III implies that the river water is suitable for fishery activities involving common species with economic value and a high tolerance to environmental conditions. Additionally, Fishery Class III indicates that the river water can be used for livestock drinking.

From the table, the average pH values of samples collected from each station and point source show minimal variation, within the range of 6.00–7.72 which indi-cates that all samples are in a neutral state. All average pH values for each station and point source are within the Class I range as per NWQS. The average dissolved oxygen (DO) concentration at each water quality station shows minimal variation, ranging from 4.30 mg/L to 4.88 mg/L. However, the average DO concentration for P1 is slightly lower at 3.86 mg/L, while the sample from P2 exhibits a higher DO concentration of 6.47 mg/L. The DO concentrations for each water sample fall within a healthy range, indicating that the amount of oxygen gas present in the water is sufficient. All DO concentrations for each station and point source fall within the Class III range as per NWQS, except for P2, which is within the Class IIA range.

The average 5-day biochemical oxygen demand (BOD5) concentration shows an ascending pattern from station W1 to station W6, ranging from 6 mg/L to 25 mg/L. The highest average BOD5 concentration is observed at station W6, reaching 25 mg/L. The water sample from P1 exhibits an average BOD5 concentration of 12 mg/L, which is relatively similar to station W3. In contrast, the average BOD5 concentration for P2 is exceptionally high, recorded at 744 mg/L. Stations W1, W2, and W3 exhibit average BOD5 concentrations within the Class III range according to NWQS. However, Station W4 and Point Source 1 have average BOD5 concentrations within the Class IV range, while Station W5, W6 and Point Source 2 have average BOD5 concentrations within the Class V range. A higher BOD5 concentration signifies the presence of an elevated concentration of organic matter in the water sample which requires more oxygen for decomposition by microorganisms. Consequently, when the BOD5 concentration is high, it may lead to oxygen deficiency in the water sample, as a significant portion of oxygen is utilized in the decomposition process. Emphasis should be directed towards monitoring and managing BOD5 concentrations at Station W4, W5, and W6 to ensure that they remain below the Class III standard outlined in the NWQS.

The average chemical oxygen demand (COD) concentrations exhibit minimal variation, ranging from 16 mg/L to 29 mg/L. The highest average COD concentration is observed at Station W5 with a value of 29 mg/L, followed by Station W6 with a value of 25 mg/L. Conversely, Station W1, W2, W3, and W4 display COD concentrations within the range of 16 mg/L to 18 mg/L. The COD concentrations for all water quality stations are within the Class IIA range according to NWQS. The average COD concentration of the point source, P1, is higher than the maximum COD concentration detected in all stations, which is 42 mg/L. P2 exhibits an exceptionally high value of 744 mg/L. Consequently, as per NWQS, the average COD concentration for P1 falls within the range of Class III, while P2 falls within the range of Class V. A higher COD concentration signifies the presence of elevated concentrations of both organic and inorganic matter in the water that can undergo chemical oxidation in the presence of oxygen. Elevated COD concentrations indicate a higher load of pollutants in the water, necessitating more oxygen for the chemical decomposition process. This increased demand for oxygen may lead to conditions of oxygen deficiency, negatively impacting aquatic ecosystems.

The average concentration of total phosphorus shows an increasing trend from station W1 to W6, ranging from 0.30 mg/L to 1.09 mg/L. Both station W1 and W2 exhibit similar total phosphorus concentrations of 0.30 mg/L and 0.35 mg/L, respectively. Station W3 has a total phosphorus concentration of 0.50 mg/L. Stations W4, W5, and W6 share similar total phosphorus concentrations of 0.90 mg/L, 0.95 mg/L, and 1.09 mg/L, respectively. All total phosphorus concentrations for each water quality station meet the Class IV standards as per NWQS. The total phosphorus concentration for P1 is 0.82 mg/L, which is relatively similar to the total phosphorus concentration for station W4. However, P2 has the highest total phosphorus concentration among all the samples, with a result of 6.10 mg/L. The total phosphorus concentrations for both P1 and P2 also satisfy the Class IV standards as per NWQS. Phosphorus is a source of nutrient for the growth of many organisms including crops and algae. High concentration of phosphorus may lead to lead to eutrophication and algae bloom. The source of phosphorus could be the agricultural runoff and wastewater discharges. Phosphorus serves as a nutrient essential for the growth of various organisms, including crops and algae. However, elevated concentrations of phosphorus can contribute to eutrophication and the occurrence of algal blooms. Common sources of phosphorus in water bodies include agricultural runoff and wastewater discharges.

The average ammoniacal nitrogen (NH3-N) concentration for each station dis-plays minimal variation, falling within the range of 0.01 mg/L to 0.03 mg/L. The average NH3-N concentrations exhibit an increasing trend from station W1 to W6. All average NH3-N concentrations for each water quality station adhere to the Class I standard as per NWQS. The average NH3-N concentrations for both P1 and P2 surpass the maximum concentration recorded among all stations, registering at 0.27 mg/L and 0.90 mg/L, respectively. Despite being higher, both NH3-N concentrations for P1 and P2 still fall within the Class I standard as per NWQS. The concentration of NH3-N in water is a valuable indicator for assessing water quality as it reflects the levels of ammonia (NH3) and ammonium ions (NH4+). Elevated concentrations of ammonia can signify a deterioration in water quality, negatively impacting the aquatic environment. High levels of ammonia can be detrimental to aquatic organisms, potentially causing harm to fish and other aquatic life.

The average total suspended solids (TSS) concentrations for stations W1, W2, and W3 range from 14 mg/L to 23 mg/L, falling within the Class I standard ac-cording to NWQS. Conversely, stations W4, W5, and W6 exhibit TSS concentrations within the range of 41 mg/L to 45 mg/L, meeting the Class IIA standard as per NWQS. P1 has the lowest TSS concentration among all the samples at 4 mg/L, aligning with the Class I standard as per NWQS. However, P2 records the highest TSS concentration at 246 mg/L, placing it within the Class IV standard as per NWQS. The TSS concentration represents the concentration of suspended solids that do not dissolve in water, including soil particles and organic matter. Elevated TSS concentrations can result in low water turbidity, affecting aquatic ecosystems by reducing light penetration into the water. This decrease in light availability may have adverse effects on the growth of aquatic plants, photosynthesis, and overall ecological balance in the aquatic environment.

Overall, stations W1, W2, and W3 exhibit better water quality compared to W4, W5, and W6, as deduced from a comprehensive analysis of various water quality parameters. All water quality parameters for stations W1, W2, and W3 generally comply with the Class III standards, with the exception of total phosphorus. On the other hand, stations W4, W5, and W6 face challenges as two water quality parameters, total phosphorus and BOD5, do not meet the Class III standards. It is evident that the upstream and middle stream sections of Sungai Batu Ferringhi largely satisfy Class III river requirements. The potential factors contributing to the lower water quality in the downstream section could be attributed to the presence of two points pollution sources and the increased flow rate of urban runoff. Attention and concerted efforts should be directed towards improving the water quality in the downstream section of Sungai Batu Ferringhi.

4.3 Hydraulic and Geometric Simulation

Figure 4 illustrates the comparison between the simulated flow rate, as predicted by the QUAL2K model, and the observed flow rate recorded during the sampling trip. The pattern of the simulated flow rate closely mirrors that of the observed flow rate, as evidenced by a perfect R-squared value of 1. From Fig. 4, a significant increase in flow rate is evident, commencing at station W4, located 0.371 km away from the downstream, and extending through to the downstream of the river. The flow rate of the river in the upstream section exhibits minimal variation. In summary, the overall trend in water flow indicates a steady flow rate at the upstream section and a consistent increase from the middle stream section to the downstream section of Sungai Batu Ferringhi.

Figure 5 illustrates the comparison between the simulated water velocity, predicted by QUAL2K, and the observed water velocity recorded during the sampling trip. The simulated water velocity closely aligns with the observed water velocity, as indicated by a high R-squared value of 0.9986.

Based on the figure, the water velocity demonstrates a decreasing trend starting from station W2, located 0.851 km away from the downstream, and continuing through the downstream section of Sungai Batu Ferringhi. Stations W1 and W2, situated in the upstream section, exhibit higher water velocities compared to the middle stream and

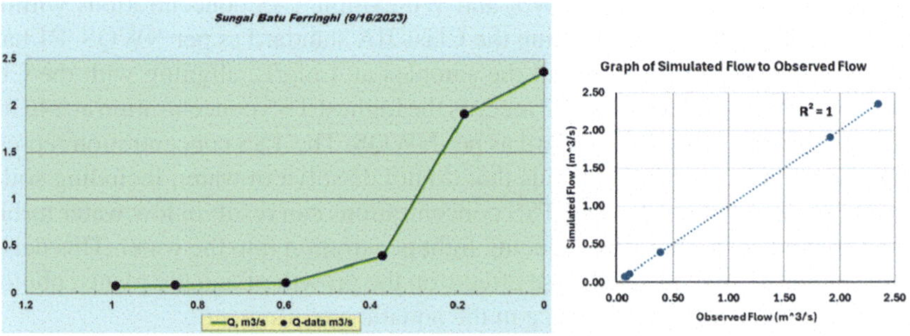

Fig. 4. Flow rate calibration.

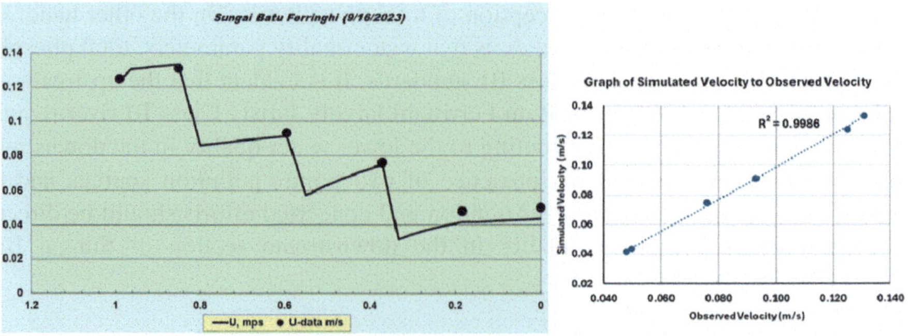

Fig. 5. Graph of water velocity calibration for Sungai Batu Ferringhi.

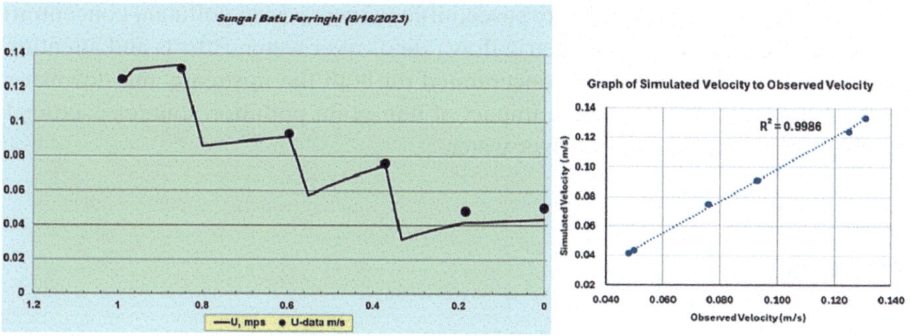

Fig. 6. Graph of water depth.

Table 6. Summary of model performance evaluation for hydraulic and geometric simulation.

Parameters	R-squared value	MAE	nRMSE	RE (%)
Flow (m^3/s)	1.0000	0.0000	0.0000	0.0000
Velocity (m/s)	0.9986	0.0032	0.0455	3.6348
Depth (m)	0.9985	0.0868	0.0649	5.6014

downstream sections. In summary, the overall trend in water velocity indicates a decline from the upper stream to the lower stream of Sungai Batu Ferringhi.

Examining Fig. 6, it is evident that the water depth at the upper stream and the middle stream, represented by stations W1, W2, W3, and W4, exhibits minimal variation. The water in these sections is characterized by shallow depths. In contrast, a substantial increase in water depth is notable at the downstream section of Sungai Batu Ferringhi. Here, the water depth surpasses 3 m, indicating a considerable depth in this downstream region.

Table 6 tabulates the results of model performance evaluation for each hydraulic and geometric parameter. The nRMSE values for all parameters are below 0.2, and the range of RE falls between 0% and 5.6%, both of which are under the 20% threshold.

The simulated hydraulic and geometric parameters exhibit a strong correlation with the observed data, accompanied by minimal statistical errors. Consequently, the simulated data is deemed acceptable.

5 Conclusion

The objective of the research was to assess the bioremediation potential of food waste The QUAL2K model has effectively simulated the water environment of Sungai Batu Ferringhi, utilizing hydrogeometric data, hydrodynamic data, water quality data, and information about point sources. By establishing a baseline model, the model has predicted the spatial water quality data of Sungai Batu Ferringhi under 7Q10 low flow conditions. As evident from the 7Q10 low flow model, the upstream and middle stream sections of Sungai Batu Ferringhi are susceptible to increases in pollutant concentration owing to the decrease in the volumetric flow rate of river water. Efforts and attention to enhance water quality should be concentrated on both the upstream and downstream sections of the river. Mitigating the impact of non-point pollution sources is crucial to preserving the over-all quality of river water.

References

1. Sudhakar, H., Subramani, T.: Study of Physico-chemical characteristics of surface water using regression analysis of Cauvery River and its tributaries in Tamilnadu, India. Asian J. Chem. **25**(6), 3199–3203 (2013)
2. Chowdhury, M.S., Othman, F., Jaafar, W.Z.W., Adham, M.I., Ema, N.: Assessment of pollution and improvement measure of water quality parameters using scenarios modeling for Sungai Selangor Basin. Sains Malaysiana. **47**(3), 457–469 (2018)
3. Abbai, S.S., Sunkad, B.N.: Effect of anthropogenic activities on zooplankton population of Sogal Pond, Belgaum District, Karnataka, India. Res. J. Recent Sci. **2**(7), 81–83 (2013)
4. Dhanalekshmy, T.G., Rajani, V., Bhaskar, M.: Water quality analysis of two surface water resources in Thiruvananthapuram, India. Res. J. Chem. Environ. **18**, 8–15 (2014)
5. Phadatare, S., Gawande, S.: Review paper on development of water quality index. IJERT. **5**(05), 765 (2016)
6. Nithyanandam, R., Huan, T., Huynh, N., Nguyen, T.T.: Case study: analysis of water quality in Sungai Batu Ferringhi. J. Eng. Sci. Technol. **10**, 15–25 (2015)
7. DOE: Environmental Quality Report 2021. Publication Section Strategic Communications Division Department of Environment Malaysia (2022)
8. USEPA: Definition and Characteristics of Low Flows (2023, August 11). https://www.epa.gov/ceam/definition-and-characteristics-low-flows
9. Rice, J., Westerhoff, P.: High levels of endocrine pollutants in US streams during low flow due to insufficient wastewater dilution. Nat. Geosci. **10**(8), 587–591 (2017)
10. Salim N. A. A.: Assessment of Water Quality: Sampling & Sample Preparation (2016). https://enviro2.doe.gov.my/ekmc/wp-con-tent/uploads/2016/08/1440479406-Top-ic%2019_Sampling&Preparation.pdf
11. Chapra, S., Pelletier, G., Tao, H.: QUAL2K: A Modeling Frame-Work for Simulating River and Stream Water Quality (Version 2.12). USA: Documentation and User's Manual (2012)

Biogas Purification Using Calcium Alginate Beads Embedded with Microbial Biomass

Mohammad Nasrun Ibrahim[1], Mohamad Anuar Kamaruddin[1] (✉), Muhammad Syahmi Najmi Adnan[1], Muhamad Izzat Samir Abdul Jalil[2], Rasyidah Alrozi[3], and Mohd Remy Rozainy Mohd Arif Zainol[4]

[1] Environmental Technology Division, School of Industrial Technology, Universiti Sains Malaysia, Gelugor, Pulau Pinang, Malaysia
anuarkamaruddin@usm.my

[2] Bioprocess Technology Division, School of Industrial Technology, Universiti Sains Malaysia, Gelugor, Pulau Pinang, Malaysia

[3] College of Engineering, UiTM Cawangan Pulau Pinang, Permatang Pauh, Pulau Pinang, Malaysia

[4] River Engineering and Urban Drainage Research Centre (REDAC), Engineering Campus, Universiti Sains Malaysia, Nibong Tebal, Pulau Pinang, Malaysia

Abstract. Biogas is a highly efficient and beneficial energy source. Biogas is an energy source generated from the anaerobic decomposition of organic materials by bacteria. The anaerobic digestion (AD) process generates biogas, which pre-sents significant economic and environmental benefits. The raw biogas pro-duced from AD is lacking in quality as it contains methane (CH_4), carbon di-oxide (CO_2), hydrogen sulfide (H_2S) and other impurities. This poses chal-lenges for its efficient utilization and requires a purification step to remove impurities. This research aims in enhancing the biogas produced by using microbial biomass embedded in calcium alginate beads. The microbial bio-mass will be taken from the AD byproduct which is the digestate. The calcium alginate beads will be prepared with a few types of ratios between sodium alginate and digestate. These beads are then used to purify the biogas by letting the gas flow through the beads. The effectiveness of the method de-pends on the changes in the concentrations of CH_4, CO_2, and H_2S in the treated gas. By eliminating the CO_2 and H_2S the percentage of CH_4 will increase, so does the quality of the biogas produced.

Keywords: Anaerobic Digestion · Biogas purification · Biogas application

A. V. Sandu et al. (Eds.): ICIR EUROINVENT 2024, SPM 64, pp. 53–64, 2025.
https://doi.org/10.1007/978-3-031-81198-2_5

1 Introduction

The challenge of energy consumption and accessibility is currently an ongoing challenge faced by a significant portion of the population in the present day. People spent a significant amount of their earnings towards acquiring fuels such as gasoline, methane, and oil. The fossil fuels are being used consistently and significantly throughout the entire process. As these forms of energy are non-renewable, their availability will decrease while their prices will increase. Hence, many are seeking alternative forms of energy. Biogas is a highly effective and economical energy source [1]. Biogas, as a renewable energy source, has played an important part in this field. Nevertheless, the process of purifying biogas continues to pose significant challenges for scientists and engineers. Biogas is regarded as a novel, environmentally friendly fuel that can be used to replace fossil fuels in various applications, including heat and steam production, electricity generation, and as a fuel for vehicles [2, 3]. Typically, it is generated in landfills or wastewater treatment plants through the anaerobic digestion of organic matter in the presence of anaerobic bacteria [4, 5]. The composition mostly consists of methane (ranging from 40% to 60%) and carbon dioxide (ranging from 40% to 55%), along with minor quantities of nitrogen-based compounds, oxygen, and water. Biogas may contain small quantities of hydrogen sulfide, halogenated compounds, and organic silicon compounds (such as siloxanes) depending on the feedstock [6]. These substances can lead to maintenance problems with equipment further down the road [7]. In order to effectively utilize biogas as a fuel, it is important to eliminate the impurities that restrict its usage as an energy source. With the exception of methane, the other gases that are found in biogas are considered to be impurities [8]. Table 1 provides an overview of the biogas that is produced from anaerobic digestion and landfill gas, breaking it down into its distinguish gas components for evaluations.

In regard to the factors mentioned above, it has become imperative to enhance the amount of CH_4 present in biogas through the process of properly separating the other gases. This is particularly important when biogas is considered to be a source of renewable energy for industrial purposes [9].

The separation requires two fundamental steps: (i) biogas cleaning/purification, which is an energy-demanding process, and refers to the removal of small unwanted components of biogas and (ii) biogas upgrading, which tries at increasing the low calorific

Table 1. Biogas composition.

Parameters	Unit	AD Biogas	Landfill gas
Methane	Vol%	60–70	35–65
Carbon Dioxide	Vol%	30–40	15–40
Nitrogen	Vol%	0–0.5	15
Oxygen	Vol%	0–1	1
Water vapour	Vol%	1–5	1–5
Hydrogen sulfide	ppm	0–4000	0–100
Ammonia	ppm	100	5
Hydrocarbon	Vol%	0	0
Siloxanne	ppm	0–41	0–100

value of the biogas and refers to the removal of CO_2 [10]. Following the separation process, the resulting biomethane is composed primarily of CH_4 (95–99%) and CO_2 (1–5%), with no detectable presence of H_2S [11]. Removing these undesired biogas components has several advantages: (i) prolonging the lifespan of process equipment, (ii) increasing the calorific value of biogas, (iii) reducing environmentally harmful emissions, and (iv) enhancing the overall economic feasibility of the process. The calorific value of biogas ranges from 20.7 to 27.8 MJ/m^3, whereas the calorific value of biomethane is higher, approximately ranging from 37.7 to 39.8 MJ/m^3 [12].

The following are the three basic categories that can be used to classify biogas purification technology: Physical methods is a wide range of methods that make use of physical separation or adsorption processes are all included. Such methods are utilized to effectively eliminate impurities that are present in the biogas stream. Chemical methods make use of chemical reactions, for the purpose of efficiently absorbing or converting pollutants that are present in biogas. Biological methods utilize the use of microorganisms, in order to facilitate the metabolic decomposition or conversion of pollutants that are present in biogas [13].

A number of physical/chemical and biological technologies have been explored for biogas purification and upgrading. Among the biological approaches, the use of immobilized microbial cells has shown promise in enhancing the productivity and efficiency of the upgrading process [14, 15]. Calcium alginate has been investigated as a matrix for the immobilization of hydrogen-producing bacteria, as it can increase the biomass density and provide a protective environment for the cells [16, 17]. This technique could potentially be applied to the removal of hydrogen sulfide and carbon dioxide from biogas as well.

The selection of the most suitable biogas purification and upgrading technology is dependent upon several criteria, including the composition of the initial biogas, the required degree of purity, the scale of the operation, and the economic and environmental concerns to be taken into account [18, 19]. Advancements in membrane technology, characterized by superior selectivity and permeability, have the potential to greatly enhance the search for the most effective biogas upgrading process [20]. This process should meet the necessary criteria for biogas utilization as car fuel and integration into the natural gas grid.

The primary advantage of using calcium alginate-immobilized microbes for biogas upgrading is the ability to maintain a high density of active biomass within a compact reactor system [21]. This can improve the overall efficiency and productivity of the purification process compared to suspended-growth systems. Additionally, the physical structure of the gel beads may provide protection for the microbial cells from inhibitory compounds in the biogas stream [22, 23].

2 Research Method

The process of creating calcium alginate beads incorporating microbial biomass requires a series of meticulous procedures to guarantee the survival of the microorganisms and the efficiency of the biocatalytic system for purifying biogas. The necessary

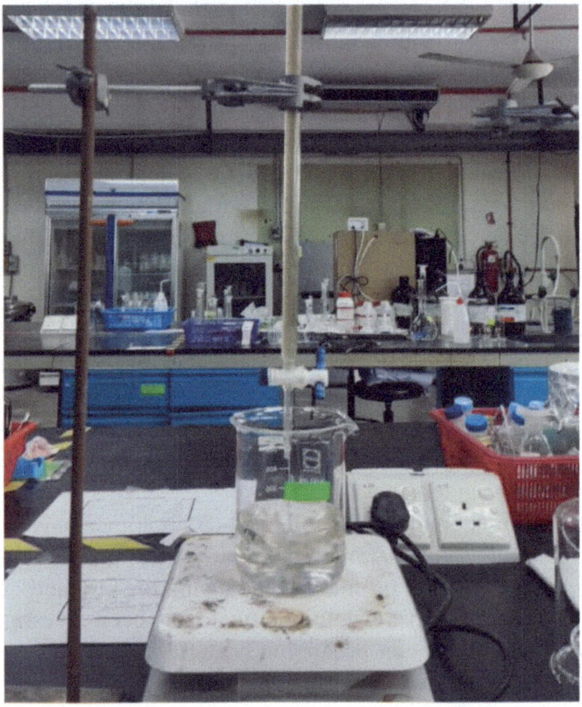

Fig. 1. Calcium alginate preparation set up.

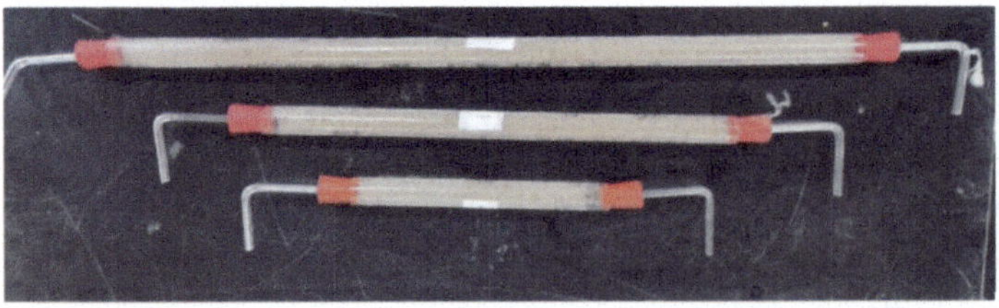

Fig. 2. Three length columns.

materials and supplies consist of sodium alginate (of commercial quality), calcium chloride ($CaCl_2$), distilled water, digestate, a magnetic stirrer, and burette beakers.

First, the sodium alginate solution must be prepared, Figs. 1 and 2. An 8 g quantity of sodium alginate powder is gradually introduced into a beaker containing 200 ml of distilled water, with constant stirring to avoid the formation of lumps. If required, the mixture is heated gradually with continuous stirring until a uniform and thick solution is achieved. Subsequently, the solution can cool down to the room temperature. Next, the solution is combined with 8 g of digestate to create a 1:1 ratio of alginate beads.

To create calcium alginate beads, start by dissolving 5 g of calcium chloride ($CaCl_2$) in 400 ml of distilled water to make a 0.1 M solution. The sodium alginate-microbial biomass mixture is dispensed in small drops into the calcium chloride solution using

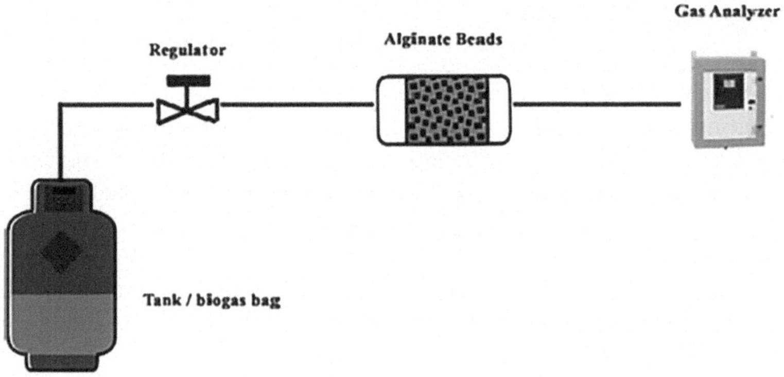

Fig. 3. Purification set up.

a burette. Upon coming into contact with the calcium chloride solution, the droplets quickly solidify, resulting in the formation of calcium alginate beads. This process continues until the entire alginate-biomass mixture is transformed into beads. The beads are after that immersed in the calcium chloride solution and left to solidify for a duration of 30–60 minutes, while being gently stirred from time to time to promote consistent hardness.

After that, the beads are thoroughly rinsed with distilled water to eliminate any excess calcium chloride. The calcium alginate beads containing microbial biomass have been produced and may now be used for biogas purification, Fig. 3. This technology is both biocompatible and efficient, effectively eliminating pollutants like hydrogen sulfide and carbon dioxide from biogas.

Subsequently, the aforementioned process is replicated utilizing alternative quantities of digestate, specifically 24 g and 40 g, in a ratio of 1:3 and 1:5, respectively, with regards to the sodium alginate to digestate ratio.

The application of calcium alginate beads embedded with microbial biomass in biogas purification involves a systematic approach to ensure optimal performance and efficiency. Firstly, the calcium alginate beads that have been produced are placed into a column. Proper packing is essential since it enables the biogas to evenly pass through the column, optimizing the interaction between the gas and the immobilized microbial cells contained in the beads. After the column is properly filled, the system is supplied with untreated biogas at a regulated rate of flow. This regulation is crucial in order to keep the residence time that is required for the biogas and the microbial biomass to interact in an effective way. The microbial cells embedded in the calcium alginate beads perform specific biochemical reactions to remove impurities such as hydrogen sulfide and carbon dioxide from the biogas. During the biogas treatment process, the released gas composition is continuously monitored to evaluate the effectiveness of contaminant removal. Specialized gas analyzers and sensors are employed to quantify the concentrations of hydrogen sulfide, carbon dioxide, and other possible impurities in the processed biogas. This data offers essential insights into the bioreactor's performance, allowing for adjustments to be made to the flow rate or operational conditions in order to enhance purification efficiency.

3 Results and Discussion

Biogas, a renewable energy source derived from organic waste, holds immense potential in mitigating greenhouse gas emissions and promoting sustainable energy practices. The purification of biogas is a critical step in maximizing its usability and environmental benefits. In recent studies, the use of calcium alginate embedded with microbial biomass has emerged as a promising approach for biogas purification. This essay delves into the analysis of biogas composition based on factors such as Beads Type, Column Length, and Retention Time, shedding light on their influence on methane (CH_4), carbon dioxide (CO_2), and hydrogen sulfide (H_2S) percentages to enhance biogas purification efficiency (Table 2).

The type of beads utilized in biogas purification methods exerts a substantial influence on the composition of biogas. The study found that the responses differed significantly depending on the type of beads used. Specifically, Response 9, which used Beads Type 1, had the lowest methane concentration at 50.48%. In contrast, Responses 1, 12, and 15, using Beads Type 3, had methane levels exceeding 51%, with Response 12 reaching a maximum of 52.2%. The significance of bead selection in maximizing methane production and enhancing biogas quality is highlighted by this variant. The length of the column is an important feature that affects both the composition of biogas and the effectiveness of its purification. The investigation showed that replies with a column length of 45 exhibited varying methane percentages, with Response 12 registering the

Table 2. Design of experiment.

		Factor 1	Factor 2	Factor 3	Response 1	Response 2	Response 3
Std	Run	A:Beads type	B:Column Length	C:Retention time	CH_4	CO_2	H_2S
			Cm	Second	%	%	ppm
11	1	3	15	300	51.8	23	1
12	2	3	45	300	52.2	23	1
14	3	3	30	180	51.22	23.02	1
1	4	1	15	180	51.2	23.03	1
17	5	3	30	180	51.22	23.02	1
8	6	5	30	300	51.82	22.98	1
2	7	5	15	180	51.27	23.01	1
4	8	5	45	180	51.28	22.99	1
5	9	1	30	60	50.48	23.04	1
13	10	3	30	180	51.22	23.02	1
15	11	3	30	180	51.22	23.02	1
3	12	1	45	180	51.23	23.02	1
9	13	3	15	60	50.48	23.04	1
6	14	5	30	60	50.5	23.04	1
7	15	1	30	300	51.79	23	1
16	16	3	30	180	51.22	23.02	1
10	17	3	45	60	50.49	23.04	1

highest value at 52.2%. Conversely, the replies that had a column length of 30 showed a reasonably stable methane proportion of approximately 51%. This discovery emphasizes the influence of column length on methane generation and emphasizes the necessity for customized column design to improve biogas purification operations. The length for which biogas components interact inside the purification system, known as retention time, has a substantial impact on the composition of biogas. Responses with a retention period of 60 seconds often showed lower methane levels compared to those with retention times of 180 or 300 seconds. Responses 1, 12, 15, and 16, all exhibited methane levels of approximately 51% continuously, with a retention time of 180 seconds. This discovery underscores the need of maximizing the duration that substances are kept in order to increase the production of methane and enhance the purity of biogas.

The efficacy of biogas purification is determined by several parameters, such as the length of the column and the type of beads used. These elements are significant in defining the composition of biogas. By examining the proportions of methane (CH_4) and carbon dioxide (CO_2) at various column lengths and with different types of beads, one can acquire useful knowledge on how to enhance the efficiency of biogas purification procedures, Fig. 4. Below are the box plots representing the data for CH_4 and CO_2.

The box plot study of CH_4 percentages across different Bead Types provides insight into the impact of bead selection on methane output in biogas purification processes. Below is a comprehensive analysis derived from the distributions displayed in the box plots.

The box plot comparing CH_4 levels between Beads Type 1 showed a narrower distribution of methane percentages, suggesting a more consistent production of methane with Beads Type 1. Nevertheless, the methane percentage for Beads Type 1 exhibited a lower median value in comparison to the other types of beads, indicating possible constraints in methane production efficiency. Beads Type 3 exhibited a box plot that

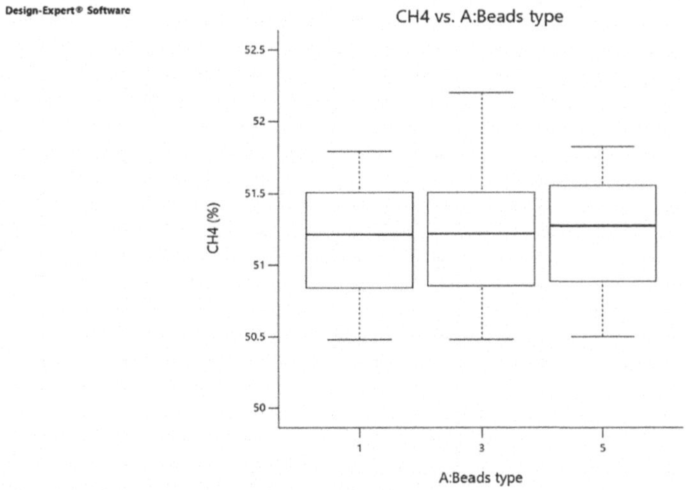

Fig. 4. Boxplot Methane Vs Beads type.

demonstrated a greater range of methane percentages compared to CH_4. Some of the data points indicated higher methane concentration. The methane percentage for Beads Type 3 was significantly higher, suggesting that it has a greater ability to produce methane compared to Beads Type 1.

Examining the box plots comparing CH_4 and Beads Type can yield valuable insights into the influence of bead selection on the efficiency of methane generation and inform methods for enhancing biogas purification procedures. The box plot study of CH_4 percentages across different Column Lengths elucidates the relationship between the length of the purification column and methane production efficiency, Fig. 5. Below is an analysis of the observations obtained from the box plots.

The box plot comparing CH_4 and Column Length 30 revealed a tight spread of methane percentages, suggesting a constant production of methane with a shorter column length. The median methane concentration for Column Length 30 remained constant, indicating effective methane production inside the system. The box plot for CH_4 versus Column Length 45 revealed a greater variation in methane percentages, with certain data points displaying higher methane content. The fluctuation in methane concentrations implies that increasing the length of the column could potentially improve methane production by allowing for a longer duration of interaction between biogas components and microbial biomass.

The box plot study of CO_2 percentages across different Beads Types provides valuable insights into the impact of bead selection on carbon dioxide levels in biogas. Below is an in-depth study derived from the distributions depicted in the box plots, Fig. 6.

The box plot comparing CO_2 levels between Beads Type 1 showed a tighter distribution, suggesting a more constant percentage of carbon dioxide with Beads Type 1. Nevertheless, the median CO_2% for Beads Type 1 exhibited a lower value in comparison to the other types of beads, indicating the possibility of variations in the efficacy

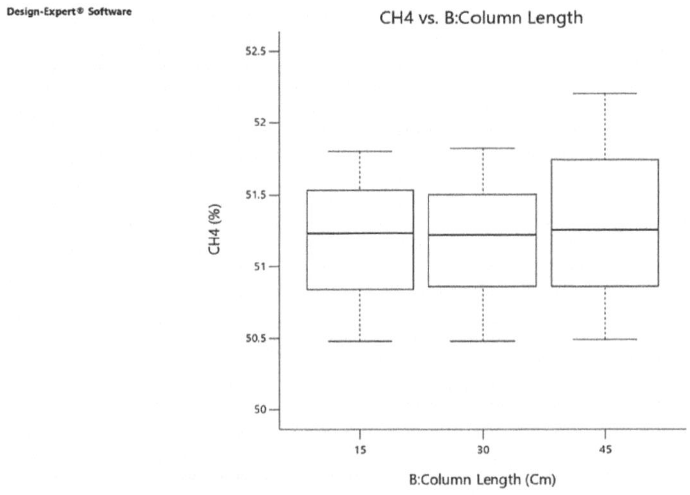

Fig. 5. Boxplot Methane Vs Column length.

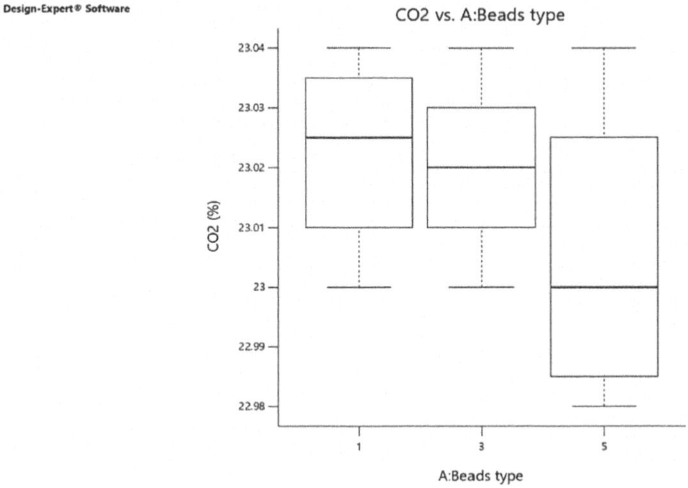

Fig. 6. Boxplot Carbon dioxide Vs Beads type.

of CO_2 removal. Beads Type 3 exhibited a box plot that demonstrated a greater range of CO_2 percentages, revealing certain responses with elevated levels of carbon dioxide. The median CO_2% for Beads Type 3 exhibited a significant increase, suggesting possible discrepancies in CO_2 elimination efficiency when compared to Beads Type 1.

Examining the box plots comparing CO_2 levels with different types of beads might yield significant information about the influence of bead selection on the efficiency of removing carbon dioxide and help in devising methods to enhance biogas purification procedures.

The box plot study of CO_2 percentages over various Column Lengths provides valuable insights into the impact of column length on carbon dioxide levels in biogas. Below is an analysis of the observations obtained from the box plots, Fig. 7.

The box plot comparing CO_2 levels with a column length of 30 showed a reasonably narrow variation of CO_2 percentages, suggesting that carbon dioxide levels remained consistent with a reduced column length.

The median CO_2 concentration for Column Length 30 remained constant, indicating effective removal of CO_2 inside the system. The box plot comparing CO_2 levels with Column Length 45 revealed a greater variation in CO_2 percentages, with certain data points displaying elevated carbon dioxide levels. The fluctuation in CO_2 percentages indicates that increasing column lengths could potentially affect the effectiveness of CO_2 removal and the purity of biogas.

Examining the box plots of CO_2 against Column Length can yield significant insights into the correlation between column design, contact duration, and carbon dioxide levels in biogas purification systems.

The above graph indicates a high degree of accuracy in the predictive model that was utilized for the purpose of calculating methane percentage. The graph exhibits a strong correlation between the actual values and the anticipated values, Fig. 8.

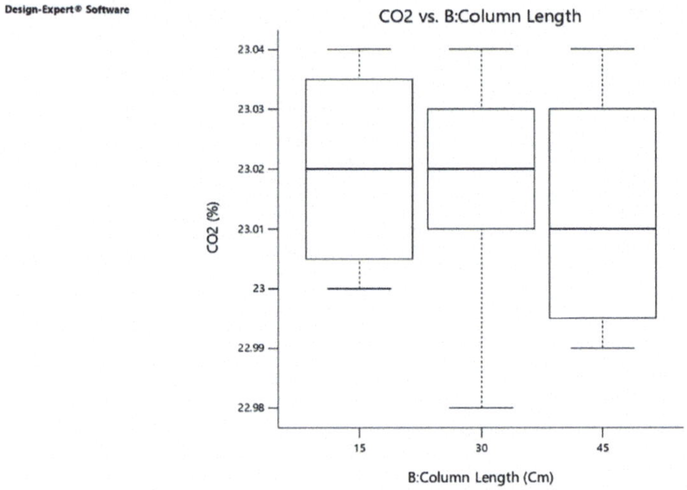

Fig. 7. Boxplot Carbon Dioxide Vs Column length.

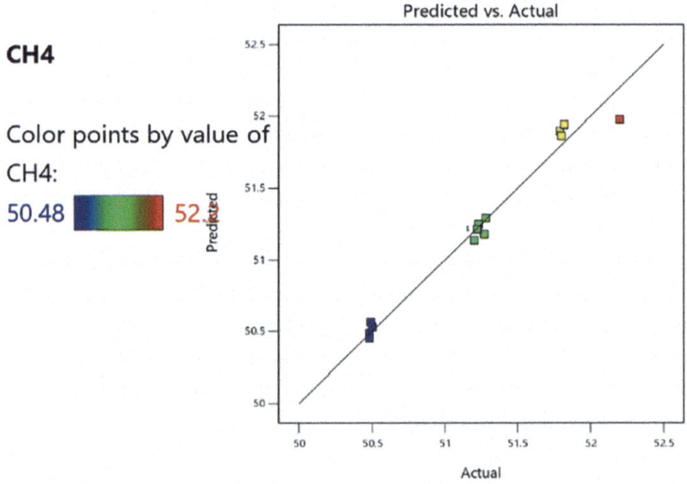

Fig. 8. Predicted Vs Actual Methane content.

The fact that the observed trends and the projected trends are consistent with one another is evidence that the model accurately captures the basic patterns and correlations that are present in the data. Therefore, the model generates reliable forecasts of the concentration of methane. The fact that the actual values and the projected values are so near to one another provides evidence that the performance of the model is impressive and trustworthy when it comes to accurately estimating methane levels.

4 Conclusion

A comprehensive review of the composition of biogas, taking into account aspects such as the type of beads, the length of the column, and the retention time, offers useful insights for enhancing the efficiency of biogas purification procedures. Researchers and practitioners can optimize biogas purification systems for improved efficiency and sustainability by comprehending the interaction of these factors on methane, carbon dioxide, and hydrogen sulfide levels. The data highlights the need of customizing bead selection, column design, and retention time optimization in order to increase methane production and enhance the quality of biogas.

Finally, an analysis of biogas composition using several criteria uncovers the complex correlation among Beads Type, Column Length, Retention Time, and the purity of biogas. By utilizing this understanding, those with a vested interest in the biogas sector can improve the methods used to remove impurities, raise the effectiveness of methane production, and make a valuable contribution towards a more environmentally friendly and long-lasting energy landscape. Additional investigation and testing are necessary to examine the interactions between these parameters and enhance the efficiency of biogas purification systems to achieve optimal environmental and economic advantages.

References

1. Xu, F., Li, Y., Ge, X., Yang, L., Li, Y.: Bioresource technology anaerobic digestion of food waste—challenges and opportunities. Bioresour. Technol. **247**, 1047–1058 (2018)
2. Zewdie, D.T., Ali, A.Y.: Cultivation of microalgae for biofuel production: coupling with sugarcane-processing factories. Energy Sustain. Soc. **10**(1), 1–16 (2020)
3. Khan, M.U., Lee, J.T.E., Bashir, M.A., Dissanayake, P.D., Ok, Y.S., Tong, Y.W., Shariati, M.A., Wu, S., Ahring, B.K.: Current status of biogas upgrading for direct biomethane use: a review. Renew. Sust. Energ. Rev. **149**, 111343 (2021)
4. Worawimut, C., Vivanpatarakij, S., Watanapa, A.: Performance evaluation of biogas upgrading systems from swine farm to biomethane production for renewable hydrogen source. Int. J. Hydrog. Energy. **44**(41), 23135–23148 (2019)
5. Abdeen, F.R.H., Mel, M., Jami, M.S., Ihsan, I., Ismail, A.F.: A review of chemical absorption of carbon dioxide for biogas upgrading. Chin. J. Chem. Eng. **24**, 693–702 (2016)
6. Khalil, M., Ali, M., Heryanto, R., Rizalie, A.: Waste to energy technology: the potential of sustainable biogas production from animal waste in Indonesia. Renew. Sust. Energ. Rev. **105**, 323–331 (2019)
7. Werkneh, A.A.: Biogas impurities: environmental and health implications, removal technologies and future perspectives. Heliyon. **8**(10), e10929 (2022)
8. Angelidaki, I., Treu, L., Tsapekos, P., Luo, G., Campanaro, S., Wenzel, H., Kougias, P.G.: Biogas upgrading and utilization: current status and perspectives. Biotechnol. Adv. **36**(2), 452–466 (2018)
9. Filarsky, F., Schmuck, C., Schultz, H.J.: Development of a biogas production and purification process using promoted gas hydrate formation – a feasibility study. Chem. Eng. Res. Des. **134**, 257–267 (2018)

10. Ilminnafik, N., Listyadi Setiyawan, D., Sutjahjono, H., Rofiq, A., Sofyan Hadi, A.: Flame characteristics of biogas from coffee waste materials. J. Phys. Conf. Ser. **1175**(1), 012273 (2019)
11. Mergenthal, M., Tawai, A., Amornraksa, S., Roddecha, S., Chuetor, S.: Methane enrichment for biogas purification using pressure swing adsorption techniques. Mater. Today Proc. **72**, 2915–2920 (2023)
12. Adnan, A.I., Ong, M.Y., Nomanbhay, S., Chew, K.W., Show, P.L.: Technologies for biogas upgrading to biomethane: a review. Bioengineering. **6**(4), 1–24 (2019)
13. San-valero, P., Penya-roja, J.M., Álvarez-hornos, F.J., Buitrón, G., Gabaldón, C., Quijano, G.: Fully aerobic bioscrubber for the desulfurization of H 2 S-rich biogas. Fuel. **241**, 884–891 (2019)
14. Angelidaki, I., Xie, L., Luo, G., Zhang, Y., Oechsner, H., Lemmer, A., Munoz, R., Kou-gias, P.G.: Biogas Upgrading: Current and Emerging Technologies, 2nd edn. Elsevier Inc. (2019)
15. Antukh, T., Lee, I., Joo, S., Kim, H.: Hydrogenotrophs-based biological biogas upgrading technologies. Front. Bioeng. Biotechnol. **10**, 1–14 (2022)
16. Damayanti, A., Kumoro, A.C., Bahlawan, Z.A.S.: Review calcium alginate beads as immobilizing matrix of functional cells: extrusion dripping method, characteristics, and application. IOP Conf. Ser. Mater. Sci. Eng. **1053**(1), 012017 (2021)
17. Giese, E.C.: Mining applications of immobilized microbial cells in an alginate matrix: an overview. Rev. Int. Contam. Ambient. **36**(3), 775–787 (2020)
18. Ahmed, S.F., Mofjur, M., Tarannum, K., Chowdhury, A.T., Rafa, N., Nuzhat, S., Ku-mar, P.S., Vo, D.V.N., Lichtfouse, E., Mahlia, T.M.I.: Biogas upgrading, economy and utilization: a review. Environ. Chem. Lett. **19**, 4137–4164 (2021)
19. Bove, D., Merello, S., Frumento, D., Al Arni, S., Aliakbarian, B., Converti, A.: A critical review of biological processes and technologies for landfill leachate treatment. Chem. Eng. Technol. **38**(12), 2115–2126 (2015)
20. Le Borgne, S., Baquerizo, G.: Microbial ecology of biofiltration units used for the desulfurization of biogas. ChemEngineering. **3**(3), 72 (2019)
21. Moestedt, J., Nordell, E., Hallin, S., Schnürer, A.: Two-stage anaerobic digestion for reduced hydrogen sulphide production. J. Chem. Technol. Biotechnol. **91**(4), 1055–1062 (2016)
22. Wang, J., Liang, J., Sun, L., Shen, J., Wang, M.: Achieving reliable partial nitrification and anammox process using polyvinyl alcohol gel beads to treat low-strength ammonia wastewater. Bioresour. Technol. **324**, 124669 (2021)
23. Pandey, S., Sarkar, S.: Anaerobic treatment of wastewater using a two-stage packed-bed reactor containing polyvinyl alcohol gel beads as biofilm carrier. J. Environ. Chem. Eng. **5**(2), 1575–1585 (2017)

Bioremediation of Refinery Oily Sludge with Food Waste Digestate as Feedstock

Muhammad Syahmi Najmi Adnan[1], Mohamad Anuar Kamaruddin[1] (✉), Mohammad Nasrun Ibrahim[1], Muhamad Izzat Samir Abdul Jalil[2], Rasyidah Alrozi[3], Norazian Mohamed Noor[4], and Mohd Remy Rozainy Mohd Arif Zainol[5]

[1] Environmental Technology Division, School of Industrial Technology, Universiti Sains Malaysia, Gelugor, Pulau Pinang, Malaysia
anuarkamaruddin@usm.my

[2] Bioprocess Technology Division, School of Industrial Technology, Universiti Sains Malaysia, Gelugor, Pulau Pinang, Malaysia

[3] Chemical Engineering Studies, College of Engineering, UiTM Cawangan Pulau Pinang, Permatang Pauh, Pulau Pinang, Malaysia

[4] Faculty of Civil Engineering & Technology, Universiti Malaysia Perlis (UniMAP), Perlis, Malaysia

[5] River Engineering and Urban Drainage Research Centre (REDAC), Universiti Sains Malaysia, Gelugor, Pulau Pinang, Malaysia

Abstract. The refinery oily sludge poses a significant challenge due to its complex composition and adverse environmental impacts. Conventional remediation methods often involve high costs and generate secondary pollutants. This study investigates the bioremediation potential of using food waste digestate as renewable resources to improve the biodegradability of refinery oily sludge. The characteristic physicochemical properties of oily (OS) sludge and food waste digestate (FWD) on pH was 5.35 and 7.52, moisture content was 6.98% and 97.22%, total solid residue was 93% and 2%, ash content 1.51% and 0.25%. Then, MLSS and MLVSS for oily sludge and food waste digestate was 164,880 mg/L, 163398 mg/L, 8440 mg/L and 7748 mg/L. The identification of bacteria in food waste digestate by Gram staining method shows that bacteria as Gram-positive, specifically of the Bacillus species. Hydrocarbon degradation was studied using gravimetric analysis. The experiment shows that higher ratios of food waste digestate result in higher degradation with 23.94% on day 11. The effect of different ratios FWD and time incubator on seed germination over a period of 21 days shows that the control condition had 3 seeds germinated, the 1:1 ratio had 4 seeds, and the 1:3 ratio had 6 seeds germinated. The highest leaf height was observed in the 1:3 ratio that reached 24.8 cm, compared to 20.3 and 22.8 cm in the control group and 1:1 ratio.

Keywords: Bioremediation · Refinery oily sludge · Food waste digestate · Hydrocarbon · Seed germination

A. V. Sandu et al. (Eds.): ICIR EUROINVENT 2024, SPM 64, pp. 65–80, 2025.
https://doi.org/10.1007/978-3-031-81198-2_6

1 Introduction

1.1 Research Background

Refinery oil sludge is a significant waste byproduct of the petroleum refining process of heavy metals, complicated mixture of hydrocarbons and various pollutants [1–5]. The digestate of food waste (FW) has higher nutrients than other digestates [6]. It is known that hydrocarbon remediation could increase soil nutrient values using food waste digestate, as it contains low metal and metalloid contents, and high nitrogen, phosphorus, and potassium contents [7]. As a result of the extensive metabolic capacities of food waste digestate, bacteria from the digestate can easily degrade various organic contaminants such as petroleum hydrocarbons [8]. In general, physical, chemical, thermal, and biological techniques can be used to remediate soil petroleum pollution [9]. Bioremediation is an environmentally friendly technique that uses selective bacterial species to remove and minimize the presence of petroleum contaminants in the environment [10]. A fundamental advantage of bioremediation lies in the synergistic interactions between the constituents of food waste digestate and petroleum oil sludge. The union of the nutrient-rich digestate and the recalcitrant oil sludge potentially activates dormant microbial communities, engendering a microbial consortium equipped to combat even the most recalcitrant hydrocarbons. Additionally, the digestate's capacity to reduces the processes that pollutants can restrict microbial activity can improve the remediation process [11–13]. By utilizing food waste digestate as soil amendment, two significant waste streams of petroleum oil sludge and food waste are synergistically transformed into resources. This decreases the challenge on landfills and exemplifies the transition from a linear "take-make-dispose" approach to a regenerative and sustainable one. By recycling the food waste digestate into renewable products such as biofertilizer, solid biofuel, and carbon-based materials shows a great potential for diverse environmental applications [14].

1.2 Problem Statement

The petroleum sectors are a vital part of modern society and the most predominant energy resource in the world. Petroleum sludge is a byproduct of refining crude oil in production plants [15]. The production of oily sludge waste has a significant volume during various industrial activities of oil treatment including exploration, mining, delivering, storing, and refining of crude oil [16, 17]. Petroleum industries post a significant pollutant defined as persistent organic pollutants (POP). These pollutants are difficult to degrade due to their low reactivity eventually pose a serious threat to both human and environmental health (USEPA) [18]. Oily sludge contains of petroleum hydrocarbons (PHCs) such as aliphatic, asphaltenes, saturated hydrocarbons, and aromatic hydrocarbons [19]. Oil sludge that is not properly managed can contaminate soil, water, and even the air, affecting the balance of precious ecosystems and posing a threat to human health due to the possibility of contaminants migrating into food sources. Various physical, chemical, or physicochemical treatment methods have been applied and introduced to treat and minimize the mineral oils sludge pollutions [20], yet the results exhibit further treatment is required. It could be mainly due to its complexity and the stability of its chemical structure, high operational cost and incomplete treatment generates a secondary pollution which required proper management [21].

1.3 Research Objectives

(i) To characteristic physiochemical of oily sludge and food waste digestate.
(ii) To identify microbial morphology of food waste digestate through the gram staining method.
(iii) To determine hydrocarbon degradation using gravimetric determination and seed germination test.

2 Literature Review

2.1 Petroleum Oily Sludge

Globally, approximately 5 tons of oily sludge is generated from every 1000 tons of crude oil [22]. Therefore, cost-effectively managing oily sludge represents a great challenge to the petroleum industry. Petroleum sludge oil is a byproduct of refining crude oil in industrial facilities [15]. The production of oily sludge waste has a significant volume during various industrial activities of oil treatment including exploration, mining, delivering, storing, and refining of crude oil [16, 17]. Total petroleum hydrocarbon (TPH) encompasses a range of hydrocarbon mixtures generated through petroleum sources, including gasoline, diesel, kerosene, and lubricating oils or greases [23]. It is imperative that these wastes are managed and disposed of properly in order to avoid the contamination of both surface and groundwater, ambient air, risk of fires, explosions, food chain contamination, and the depletion of habitats for plants [24]. The discharge of petroleum through spilling leaking, and other ways of discharge frequently leads to the pollution of both groundwater and soil, particularly in situations involving greater scales [25]. The safe disposal of these residues is the main challenges facing refineries and petrochemical industries, since its fate and improper treatment may have serious consequences for the environment and may pose a risk to public health [26–28]. Even if de-oiled, the oily sludge cannot be placed in landfills unless properly treated.

2.2 Toxicity of Oily Sludge

When oily sludge is disposed of in the environment, it can cause a variety of toxic effects in the environment that are caused by PHCs and heavy metals [29]. Heavy metals are cumulatively hazardous and pose a particular hazard. PHC are the most concerning because of their genotoxic effects on humans and other ecological organisms [10, 30]. As a result of PHC accumulation in oily sludge, they might migrate through the soil profile and enter water that connects to other aquatic systems, posing serious risks such as the reduction of fish diversity and abundance in aquatic ecosystems [31, 32]. As a result of the presence of PHCs in oily sludge, it is possible that these molecules could cause a decrease in soil enzyme activity such as hydrogenase and invertase and cause toxic effects on soil microorganisms [33, 34]. The removal of oily sludge from lagoons lined with cement and bricks still results in odor problems and fire hazards [35]. It has also been noted that refinery oily sludge disposed in lagoons or landfills can have a significant impact on atmospheric VOC [36]. Consequently, communities and workers are more likely to suffer from health problems. Numerous laws have been enacted around the world to deal with, store, and dispose of petroleum sludge due to its potential dangers, such as the RCRA [37]. The inadequate management and inappropriate disposal of petroleum sludge can have serious consequences affecting the ecosystem, human well-being, aquatic ecosystems, and the quality of soil [38, 39].

2.3 Food Waste Digestate

Food waste digestate has higher nutrients than other digestates [6]. It is known that hydrocarbon remediation could increase soil nutrient values using food waste digestate due to its diminished levels of metals and metalloids, as well as its elevated concentrations of nitrogen, phosphorus, and potassium [7]. As a result of the extensive metabolic capacities of food waste digestate, bacteria present in the digestate exhibit a high capacity for the degradation of diverse organic pollutants, including petroleum hydrocarbons [8]. This increased growth of bacteria results in an increased rate of hydrocarbon degradation at a shorter period [40]. Microorganisms produced by digestate possess the capacity to participate in the process of decomposition of petroleum products, corresponding to their counterparts coming from hydrocarbon-contaminated environments [8]. Recently, digestate has gained popularity as an organic fertilizer due to its high nutrient content and low carbon content that's easy to degrade [41]. This digestate contains a substantial proportion of organic substance with high concentration of mineral nitrogen (ammonium nitrogen [NH_4^+-N] and nitrate nitrogen [NO_3^- -N]) which makes it an ideal feedstock for agriculture and chemicals industries [42].

2.4 Bioremediation

The term "bioremediation" refers to the method of utilizing bacteria, algae, fungi, and plants to degrade, transform, eliminate, immobilize, or detoxicate several physical and chemical contaminants to the surroundings [43]. Bioremediation technology has been extensively used by experts for preserving and purifying soil pollutions includes petroleum hydrocarbons [44, 45]. Bioremediation techniques that based on bacteria and fungi, including other living microorganisms through transformation, degradation and mineralization of environmental toxicity treatment that can reduce the levels of contaminants and/or their toxicity with cost-effective [45], adaptable, and eco-friendly [46]. The biodegradation capability of the microbes involved in the bioremediation technique is affected by type, abundance, and biodegradation potential [21]. According to [47] the implementation of microbes' consortium for the removal of sophisticated contaminants such as refining oily sludge could be a better option than using separate microbes due to synergistic interactions between various consortium members [48].

Additionally, biodegradation rates are extremely influenced by physicochemical parameters for example pH, temperature, salinity, oxygen, nutrient availability, and concentration of the contaminants [49]. Microorganisms that degrade petroleum hydrocarbons utilize carbon compounds for energy, growth, and reproduction [50]. The interactions of microbes with chemicals that sustain them with energy and nutrition is essential for their success in eliminating contaminants [51]. Consequently, empirical adjustment of both biotic and abiotic parameters is required to improve efficient bioremediation approaches for the removal of petroleum hydrocarbon pollutions [52].

3 Material and Method

3.1 Sample Collection

Refinery oily sludge sample was collected from oil tanker sludge at Malaysia Marine and Heavy Engineering (MMHE) East Yard Pasir Gudang, Johor. The location at 1°26′26"N 103°55′17″E. Food waste digestate sample was collected at Pusat Komuniti Lestari (Environment Idaman) Jitra Kedah, Malaysia. The location at 6°15′24"N 100°27′16″E.

3.2 Characteristics Physiochemical of Oily Sludge (OS) and Food Waste Digestate (FWD)

The samples are stored in the fridge at 4 °C until subsequent analysis.

3.3 Characteristics Physiochemical of OS and FWD

pH The pH of the sample was measured using a pH meter. The pH meter was calibrated by pressing "CAL" while pressing "READ". Diluted 40 mL hexane solution into OS before measuring the pH value Then, a pH probe was placed in the sample and the value was recorded when the display was stable. Hexane was chosen because it is an organic solvent and would not affect the pH of the oily sludge sample [53].

Moisture Content and Total Solid Residue (TSR) The moisture content of OS was measured using the oven-drying method following the ASTM D-3173 standard guidelines. Three crucibles were cleaned, dried, and weighed and the values were labelled as W1. The OS was poured into crucibles and weighed, and the values obtained were designated as W2. The crucibles were placed in an oven and dried at a temperature of 110 °C. After 24 h, the crucibles were removed from the oven and cooled in a desiccator; the crucibles were weighed again and denoted as W3. The moisture content and total solid residue (TSR) were calculated using the equations below:

$$\text{Moisture content},\% = \frac{(W2 - W1) - (W3 - W1)}{(W2 - W1)} \times 100 \quad (1)$$

$$\text{Total Solid Residue}\left(\text{TSR}\right),\% = \frac{\left(\text{W3}-\text{W1}\right)}{\left(\text{W2}-\text{W1}\right)}\times 100 \tag{2}$$

W1 = weight of crucible; W2 = weight of crucible with oily sludge before oven drying; W3 = weight of crucible with oily sludge after oven drying.

Ash Content The clean crucible was placed in an oven at 105 °C for 20 minutes and then cooled down by inserting it into a desiccator. Next, the crucible was weighed without a cap and the weight was recorded. The sample was then added to the crucible, and the combined weight was recorded. The crucible with the sample was then placed in a muffle furnace at 500 °C for 40 minutes, after which it was cooled in a desiccator for 10 minutes. Finally, the crucible with the sample was weighed again to 4 decimal places, and the percentage of ash was calculated by comparing the initial and final weights. The calculated percentage of ash was calculated using the equations below:

$$\text{Ash Content},\% = \frac{\left(\text{W2}-\text{W1}\right)}{\text{W3}}\times 100 \tag{3}$$

W1 = Weigh crucible with no cap; W2 = Weigh crucible with sample; W3 = Weigh crucible with ash.

MLSS The weight of the empty pan with a glass-fiber filter paper was measured using a weighing apparatus. A 25 mL sample was filtered through a glass-fiber filter paper and heated at 105 °C for 1 hour. Then, it was transferred to a desiccator for 15 minutes to balance the temperature and weight, accordingly. The calculation for MLSS is shown below:

$$\text{MLSS, mg / L} = \frac{\left(\text{W2}-\text{W1}\right)}{\text{sample volume, mL}}\times 1000 \tag{4}$$

W1 = weight of the glass-fiber filter paper; W2 = weight of the glass-fiber filter paper with dried residue in mg.

MLVSS The pan with the glass-fiber filter paper (Whatman grade GF/A, 1.6 μm) and residue after the MLSS analysis was heated in a furnace at 550 °C for 20 minutes. Then, it was transferred to the desiccator to balance the temperature and weight, accordingly. The calculation for MLVSS is shown below:

$$\text{MLVSS, mg / L} = \frac{\left(\text{MLSS}-\text{W3}\right)}{\text{sample volume, mL}}\times 1000 \tag{5}$$

where W3 is the weight of the glass-fiber filter paper with dried residue in mg.

3.4 Culture Media Preparation

Prepare 28 g of nutrient agar powder in 1 L in distilled water into 1 L of schott bottle. Then shaking frequently until completely dissolved and sterilize in autoclaved at 121 °C, 15 psi pressure for 15 min so that no unwanted microorganisms grow which may contaminate the growing sample. This is because the nutrient agar media contained essential nutrients for the growth of the microorganisms. The melted sterile agar was poured quickly into a series of sterilize petri dishes. The petri dish was filled to about one third capacity with melted agar. Pour the liquid into petri dish and waited for the medium to solidify. The procedure undergoes in laminar flow cabinet.

3.5 Sample Culture by Serial Dilution and Plating Method

Prepare FWD sample and culture by serial dilution and plating method following [54]. This process helps to reduce the concentration of microorganisms in the sample, making it easier to count the colonies on an agar plate. Spread Plate this technique involves spreading a small number of bacteria suspended in a solution over the surface of an agar plate using a sterile L-shape glass rod (Spreader). The procedure undergoes in laminar flow cabinet.

3.6 Gram Staining

Gram-staining following method by [55]. Gram staining is a crucial technique in microbiology used to classify bacteria into two main groups which are gram-positive and gram-negative. The key points about gram staining are differentiates bacteria based on the chemical and physical properties of their cell walls, Gram-positive cells have a thick layer of peptidoglycan that retains the primary crystal violet stain, and Gram-negative cells have a thinner peptidoglycan layer that allows the crystal violet to be washed out by the decolorizing agent.

3.7 Experimental Design

The determine hydrocarbon degradation potential of OS and/or FWD were incubator with several days. The composition of modified Bushnell-Haas medium broth was as follows: ($MgSO_4$(0.2), $CaCl_2$(0.02), KH_2PO_4(1.0), $(NH_4)_2HPO_4$(1.0), NH_4Cl(3.5) and $FeCl^3$(0.05) gL^{-1}. While for of nutrient broth composition was (Beef extract (1.0), peptone (5.0), yeast extract (2.0) and sodium chloride (5.0) gL^{-1}. Both broths were shaking frequently until completely dissolved and sterilize in autoclaved at 121 °C, 15 psi pressure for 15 min. The experiment was presented in Table 1 using 250 ml Erlenmeyer flask with 100 ml of modified Bushnell-Haas broth medium [8], 20 ml of nutrient broth, 1 ml of oily sludge and 1 ml and/or 3 ml of food waste digestate.

Table 1. Composition of enrichments from different ratio of FWD.

No.	Enrichment condition
I	BH (100 mL) + NB (20 mL) + OS (1 mL)
II	BH (100 mL) + NB (20 mL) + OS (1 mL) + FWD (1 mL)
III	BH (100 mL) + NB (20 mL) + OS (1 mL) + FWD (3 mL)

The (I) as the control condition and five experimental conditions each with a different time of incubation. The sample pH was adjusted to 7 and sample condition is set at 150 rpm, 37 °C for 1,3,5,9 and 11 days of rotary shaker incubator.

3.8 Biodegradation Study by Gravimetric Analysis

The process of breaking down hydrocarbons was examined by gravimetric analysis. After incubating for different durations at a temperature of 37 °C, the 250 ml flasks were taken out and bacterial activity ended by adding 1% (v/v) 1 N-HCL. To separate the remaining oil, a mixture of 40 ml of culture broth, petroleum ether, and acetone in a 1:1 ratio was vigorously combined in a separating funnel, resulting in the formation of a single emulsified layer. Acetone was introduced and vigorously mixed to break down the emulsification, which resulted in the formation of 3 separate layers. The uppermost stratum is comprised of petroleum ether, oily sludge, and acetone; the middle layer consists of clumping cells, and the lower aqueous layer contains acetone and water in a soluble state. The lower two layers were dispersed, while the upper layer, consisting of petroleum ether mixed with oil and acetone, was transferred into a pre-weighed, clean beaker. This procedure was repeated 2–3 times to remove all remaining oil. To remove moisture from the oil that was produced, it was filtered via anhydrous sodium sulphate. The petroleum ether and acetone layer were evaporated on a water bath and allowed to vaporize in a fume hood at a temperature of 90 °C for 2 hours. Subsequently, the beaker was extracted, allowed to cool at room temperature for a duration of 15 minutes, and then measured to determine the remaining amount of oil. The gravimetric test measured the weight of residual oil remaining after biodegradation and compared it to the initial amount of oil in the beaker. Equation used is shown below:

$$\text{Hydrocarbon degradation},\% = \frac{\text{Wroc} - \text{Wros}}{\text{Wroc}} \times 100 \tag{6}$$

Wroc = weight of residual oil in abiotic control; Wros = the weight of residual oil in test sample in gram.

3.9 Seed Germination Test

Prepare 15 samples ratio of OS and FWD with uncontaminated soil/natural soil (150 g) with different incubator times presented in Table 2. Each pot will be planted with 6 seeds of mung bean. This experiment was to determine the effect different ratio FWD and incubation time on seed germination mixture.

Table 2. Condition of OS and FWD on seed germination growth.

No.	Enrichment condition
I	BH (100 mL) + NB (20 mL) + OS (1 mL) + soil (150 g)
II	BH (100 mL) + NB (20 mL) + OS (1 mL) + FWD (1 mL) + soil (150 g)
III	BH (100 mL) + NB (20 mL) + OS (1 mL) + FWD (3 mL) + soil (150 g)

Table 3. Characteristic physiochemical of OS and FWD.

Parameter	Unit	Oily sludge	Food waste digestate
pH	–	5.35	7.52
Moisture content	%	6.98	97.22
Total solid residue, TSR	%	93	2
Ash content	%	1.51	0.25
MLSS	mg/L	164,880	8440
MLVSS	mg/L	163,398	7748

4 Result and Discussion

4.1 Characteristic Physiochemical of OS and FWD

OS has an acidic characteristic with pH value of 5.35 while FWD has a neutral to slightly alkaline pH value of 7.52. Oily sludge is relatively dry with a moisture content of 6.98% while FWD has high moisture of 97.22% nearly a liquid, Table 3.

Thus, for TSR the OS consists of almost 93% solid while the FWD consist of only 2% solids. For ash content for OS has 1.51% ash content while FWD has 0.25%, this implies more inorganic materials within the sludge. MLSS and MLVSS for OS has significantly higher values for both MLSS (164,880 mg/L) and MLVSS (163,398 mg/L) compared to FWD the MLSS value was 8440 mg/L VS, 7748 mg/L MLVSS, which reveal that the OS has much higher concentration of suspended solids, both organic and inorganic.

4.2 Identification of Morphology Bacteria in FWD Using Gram Staining Method

On the basis of the Gram staining technique bacteria isolated is found to be Gram positive and belonging to Bacillus species, Fig. 1. The organism is grossly morphologically typical of elongated rods palisade arrangement and has classic gram-positive characters.

This structure which has a complicated layer of peptidoglycan enamouring the cell makes the cell to be rigid and elastic ensuring survival in different environments as well as versatility.

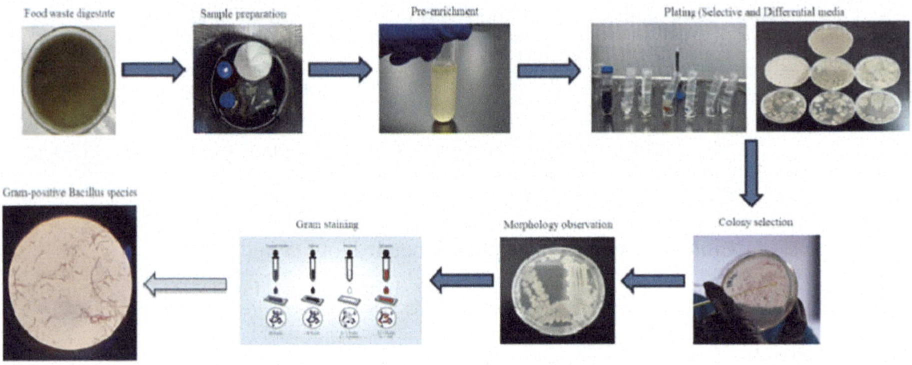

Fig. 1. The process of identification of gram staining method.

4.3 Hydrocarbon Degradation Potential of OS with Different Time Incubator and FWD Ratio

The degradation rate measured by the percentage decrease in weight from the initial to final weight, increases as the ratio of the incubation mixture increases from 1: 1 to 1: 3, Fig. 2. At the 1:1 ratio, the degradation rate ranges from 4.84% to 15.03% across the various incubation days. The 1:3, indicating that the degradation rate increase between 8% and 9%. 16% to 23. 94% across the incubation days of the different samples. As the incubation days increase from 1 to 11 days, the degradation rates generally increase for both the 1:1 and 1:3 ratios. Overall, the data suggests that increasing the ratio of the incubation mixture from 1:1 to 1:3 showed higher degradation rate.

These findings imply that the ratio of food waste digestate and the period of incubation affect the degradation rate of petroleum hydrocarbons, Fig. 3.

4.4 The Effect of Different Day of Incubator and Ratios of OS and FWD on Seed Germination

The data shows the impact of varying proportions of OS and FWD on the ability to germinate seed for 21 days, Fig. 4.

The ratios tested were 1:1 and 1:3 (OS: FWD) with a control group of 0 (No FWD). At day 1, the control condition and 1:1 ratio had 0 seeds germinated while the 1:3 ratio had 3 seeds germinated. By day 21, the control condition had 3 seeds germinated, 1:1 ratio, 4 seeds germinated, while in the 1:3 ratio, six seeds germinated. The highest leaf height was observed in the 1:3 ratio that reaching 24.8 cm. The table showed that the number of seeds planted with food waste digestate seems to have some influence on the plant with higher numbers of seeds 1:3.

The results showed that a ratio of OS to FWD of 3 had a positive impact on seed germination and growth with more seeds germinated and taller height of the leaves than of the control group, Fig. 5.

Fig. 2. The experiment condition of OS with FWD digestate ratio.

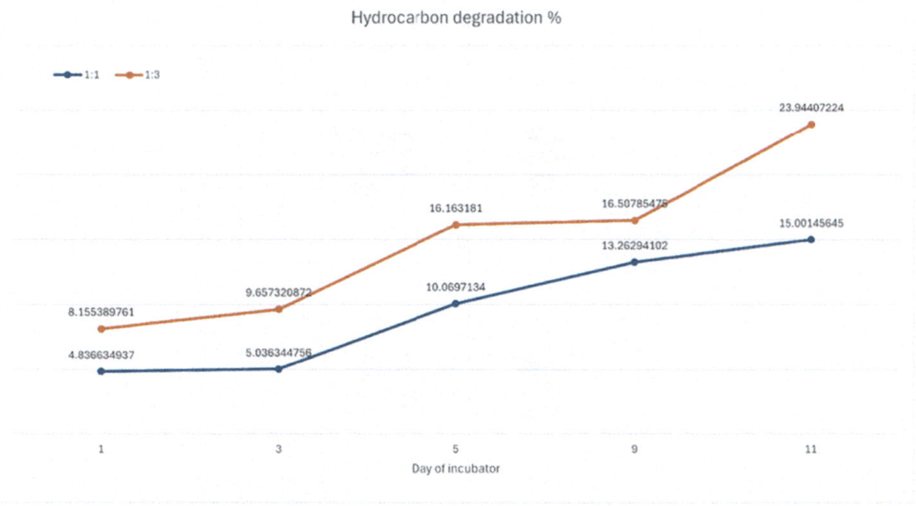

Fig. 3. Hydrocarbon degradation over time.

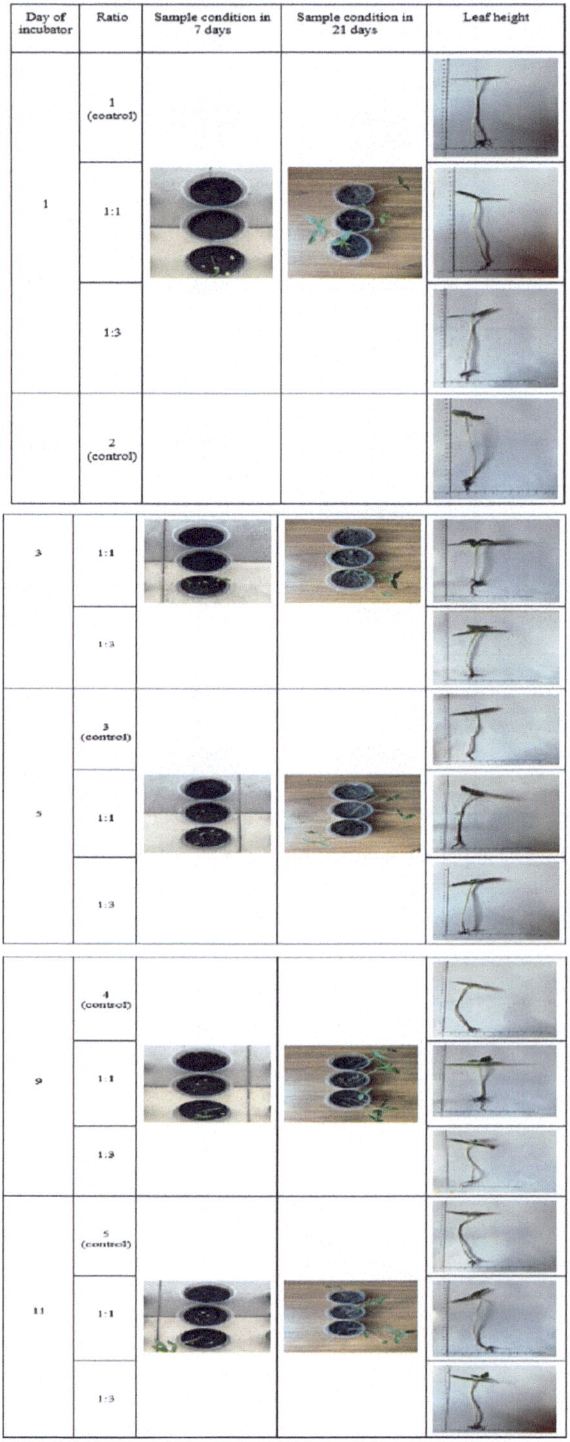

Fig. 4. Effect of different day of incubator and ratios of OS and FWD on seed germination.

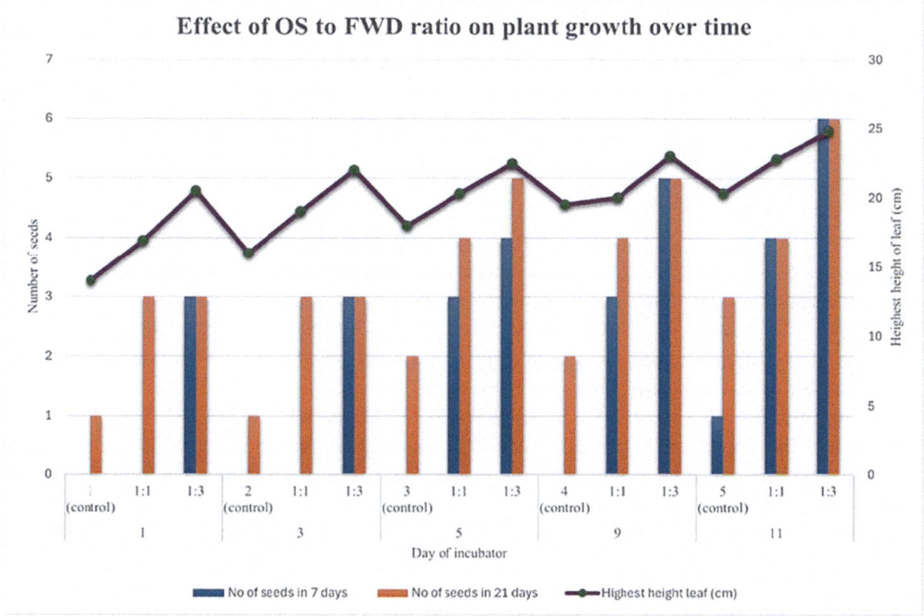

Fig. 5. Effect of oily sludge to FWD ratio on plant growth over time.

5 Conclusion

The objective of the research was to assess the bioremediation potential of food waste digestate on refinery oily sludge. Microbial characterization of the food waste digestate through microscopy revealed the presence of Gram-positive Bacillus species bacteria. The hydrocarbon degradation study using gravimetric analysis showed that higher ratios of food waste digestate (1:3) contribute to higher degradation percent than the control that incubated for 11 days. The seed germination test showed that there is an enhancement in plant growth after the application of oily sludge mixed with food waste digestate where the plant with higher concentration of food waste digestate showed higher leaf height than the control. In conclusion, bioremediation process can utilize microorganisms to degrade pollutants offers a promising alternative for the treatment of refinery oily sludge.

References

1. Niu, A., Sun, X., Lin, C.: Trend in research on characterization, environmental impacts and treatment of oily sludge: a systematic review. Molecules. **27**, 1–23 (2022)
2. Sabour, M.R., Besharati, M., Dezvareh, G.A., Hajbabaie, M.R., Akbari, M.: Application of artificial neural network with the back-propagation algorithm for estimating the amount of polycyclic aromatic hydrocarbons in Tehran Oil Refinery, Iran. Environ. Nanotechnol. Monit. Manag. **18**, 100677 (2022)

3. Tao, X., Ma, L.A.: Bioremediation of refinery oil contaminated soil. OAlib. **5**, 1–6 (2018)
4. Khatatbeh, M., Alzoubi, K., Khabour, O., Al-Delaimy, W.: Adverse health impacts of living near an oil refinery in Jordan. Environ Health Insight. **14**, 1178630220985794 (2020)
5. Singh, B., Kumar, P.: Physicochemical characteristics of hazardous sludge from effluent treatment plant of petroleum refinery as feedstock for thermochemical processes. J. Environ. Chem. Eng. **8**, 103817 (2020)
6. Opatokun, S.A., Strezov, V., Kan, T.: Product based evaluation of pyrolysis of food waste and its digestate. Energy. **92**, 349–354 (2015)
7. Peng, W., Pivato, A.: Sustainable management of digestate from the organ-ic fraction of municipal solid waste and food waste under the concepts of back to Earth alternatives and circular economy. Waste Biomass Valorization. **10**, 465–481 (2019)
8. Gielnik, A., Pechaud, Y., Huguenot, D., Cebron, A., Esposito, G., van Hullebusch, E.D.: Bacterial seeding potential of digestate in bioremediation of diesel contaminated soil. Int. Biodeterior. Biodegradation. **143**, 104715 (2019)
9. Kuppusamy, S., Thavamani, P., Venkateswarlu, K., Lee, Y.B., Naidu, R., Megharaj, M.: Remediation approaches for polycyclic aromatic hydrocarbons (PAHs) contaminated soils: technological constraints, emerging trends and future directions. Chemosphere. **168**, 944–968 (2017)
10. Patel, A.B., Shaikh, S., Jain, K.R., Desai, C., Madamwar, D.: Polycyclic aromatic hydrocarbons: sources, toxicity, and remediation approaches. Front. Microbiol. **11**, 2675 (2020)
11. Selvaraj, P.S., Periasamy, K., Suganya, K., Ramadass, K., Muthusamy, S., Ramesh, P., Bush, R., Vincent, S.G.T., Palanisami, T.: Novel resources recovery from anaerobic digestates: current trends and future perspectives. Crit. Rev. Environ. Sci. Technol. **52**, 1915–1999 (2022)
12. Oghoje, S., Ukpebor, J., Ukpebor, E.: The effects of chicken manure digestates on the removal of diesel range organics from petroleum products polluted soils S. Bulg. J. Soil Sci. **6**, 78–95 (2021)
13. Song, S., Lim, J.W., Lee, J.T.E., Cheong, J.C., Hoy, S.H., Hu, Q., Tan, J.K.N., Chiam, Z., Arora, S., Lum, T.Q.H., Lim, E.Y., Wang, C.H., Tan, H.T.W., Tong, Y.W.: Food-waste anaerobic digestate as a fertilizer: the agronomic properties of untreated digestate and biochar-filtered digestate residue. Waste Manag. **136**, 143–152 (2021)
14. Alberto, D. R.: Sustainable food waste management through the integration of anaerobic digestion and thermochemical processes. Thesis. Rochester Institute of Technology (2022)
15. Thong, O., Ibrahim, A., Rahim, H.A., Yong, W.J., Ismail, N.: A review of crude oil recovery methods from petroleum sludge. Int. J. Eng. Technol. Sci. **8**(1), 40–50 (2021)
16. Hu, G., Li, J., Hou, H.: A combination of solvent extraction and freeze thaw for oil recovery from petroleum refinery wastewater treatment pond sludge. J. Hazard. Mater. **283**, 832–840 (2015)
17. Zhao, Y., Yan, X., Zhou, J., Li, R., Yang, S., Wang, B., Deng, R.: Treatment of oily sludge by two-stage wet air oxidation. J. Energy Inst. **92**, 1451–1457 (2019)
18. US EPA: Persistent Organic Pollutants: A Global Issue, A Global Response. https://www.epa.gov/international-cooperation/persistent-organic-pollutants-global-issue-global-response. Accessed 6 Jul 2023
19. Ma, Y., Yao, M., Liu, L., Qin, C., Qin, B., Deng, N., Liang, C., Yao, S.: Mechanism and characteristics of oil recovery from oily sludge by sodium lignosulfonate treatment. ACS Omega. **6**, 25819–25827 (2021)
20. Johnson, O.A., Affam, A.C.: Petroleum sludge treatment and disposal techniques: a review. Environ. Sci. Pollut. Res. **29**, 40358–40372 (2022)
21. Varjani, S.J.: Microbial degradation of petroleum hydrocarbons. Bioresour. Technol. **223**, 277–286 (2017)

22. Teng, Q., Zhang, D., Yang, C.: A review of the application of different treatment processes for oily sludge. Environ. Sci. Pollut. Res. **28**, 121–132 (2021)
23. Adedeji, J.A., Tetteh, E.K., Opoku Amankwa, M., Asante-Sackey, D., Ofori-Frimpong, S., Armah, E.K., Rathilal, S., Mohammadi, A.H., Chetty, M.: Microbial bioremediation and biodegradation of petroleum products—a mini review. Appl. Sci. Switz. **12**, 12212 (2022)
24. Islam, B.: Petroleum sludge, its treatment and disposal: a review. Int. J. Chem. Sci. **13**, 1584–1602 (2015)
25. Xia, W., Du, Z., Cui, Q., Dong, H., Wang, F., He, P., Tang, Y.C.: Biosurfactant produced by novel Pseudomonas sp. WJ6 with biodegradation of n-alkanes and polycyclic aromatic hydrocarbons. J. Hazard. Mater. **276**, 489–498 (2014)
26. Almansoory, A.F., Hasan, H.A., Abdullah, S.R.S., Idris, M., Anuar, N., Al-Adiwish, W.M.: Biosurfactant produced by the hydrocarbon-degrading bacteria: characterization, activity and applications in removing TPH from contaminated soil. Environ. Technol. Innov. **14**, 100347 (2019)
27. Cerqueira, V.S., Peralba, M.d.C.R., Camargo, F.A.O., Bento, F.M.: Comparison of bioremediation strategies for soil impacted with petrochemical oily sludge. Int. Biodeterior. Biodegradation. **95**, 338–345 (2014)
28. Rodriguez-Campos, J., Perales-Garcia, A., Hernandez-Carballo, J., Martinez-Rabelo, F., Hernández-Castellanos, B., Barois, I., Contreras-Ramos, S.M.: Bioremediation of soil contaminated by hydrocarbons with the combination of three technologies: bioaugmentation, phytoremediation, and vermiremediation. J. Soils Sediments. **19**, 1981–1994 (2019)
29. Sinha, B., Roy, S., Kumar, K.: Bioremediation of oily sludge: a case base analysis to sustainable supply chain. Res. Environ. Sustain. **2**, 100008 (2020)
30. Robertson, S.J., McGill, W.B., Massicotte, H.B., Rutherford, P.M.: Petroleum hydrocarbon contamination in boreal forest soils: a mycorrhizal ecosystems perspective. Biol. Rev. Camb. Philos. Soc. **82**, 213–240 (2007)
31. Wake, H.: Oil refineries: a review of their ecological impacts on the aquatic environment. Estuar. Coast. Shelf Sci. **62**, 131–140 (2005)
32. Honda, M., Suzuki, N.: Toxicities of polycyclic aromatic hydrocarbons for aquatic animals. Int. J. Environ. Res. Public Health. **17**, 1363 (2020)
33. Suleimanov, R.R., Gabbasova, I.M., Sitdikov, R.N.: Changes in the properties of oily gray forest soil during biological reclamation. Izvestiia Akademii nauk Seriia biologicheskaia/ Rossiiskaia akademiia nauk. **32**, 109–115 (2005)
34. Koolivand, A., Abtahi, H., Parhamfar, M., Saeedi, R., Coulon, F., Kumar, V., Villaseñor, J., Sartaj, M., Najarian, N., Shahsavari, M., Seyedmoradi, P., Rahimi, L., Bagheri, F.: The effect of petroleum hy-drocarbons concentration on competition between oil-degrading bacteria and indigenous compost microorganisms in petroleum sludge bioremediation. Environ. Technol. Innov. **26**, 102319 (2022)
35. Bhattacharyya, J.K., Shekdar, A.V.: Treatment and disposal of refinery sludges: Indian scenario. Waste Manag. Res. **21**, 249–261 (2003)
36. Suganya, K., Sivapragasam, C., Sharma, N.K., Vanitha, S.: Current trends on oil sludge characterization, toxicity and treatment systems. Int. J. Recent Technol. Eng. **8**, 13–17 (2019)
37. Hu, G., Li, J., Zeng, G.: Recent development in the treatment of oily sludge from petroleum industry: A review. J. Hazard. Mater. **261**, 470–490 (2013)
38. Cheng, S., Chang, F., Zhang, F., Huang, T., Yoshikawa, K., Zhang, H.: Progress in thermal analysis studies on the pyrolysis process of oil sludge. Thermochim. Acta. **663**, 125–136 (2018)
39. Saleh, H.M., Moussa, H.R., Mahmoud, H.H., El-Saied, F.A., Dawoud, M., Wahed, R.S.A.: Potential of the submerged plant Myriophyllum spicatum for treatment of aquatic environments contaminated with stable or radioactive cobalt and cesium. Prog. Nucl. Energy. **118**, 103147 (2020)

40. Liu, S., Sun, R., Cai, M., Kong, Y., Gao, Y., Zhang, T., Xiao, X., Qiao, Y., Xue, J., Huang, G.: Petroleum spill bioremediation by an indigenous constructed bacterial consortium in marine environments. Ecotoxicol. Environ. Saf. **241**, 113769 (2022)
41. Kataki, S., Hazarika, S., Baruah, D.C.: Investigation on by-products of bioenergy systems (anaerobic digestion and gasification) as potential crop nutrient using FTIR, XRD, SEM analysis and phyto-toxicity test. J. Environ. Manag. **196**, 201–216 (2017)
42. Makádi, M., Tomócsik, A., Orosz, V.: Digestate: a new nutrient source – review. In: Kumar, S. (ed.) Biogas, pp. 295–310. In Tech (2012)
43. Gouma, S., Fragoeiro, S., Bastos, A.C., Magan, N.: Bacterial and fungal bioremediation strategies. In: Microbial Biodegradation and Bioremediation, pp. 302–323. Elsevier (2014)
44. Sonawane, J.M., Rai, A.K., Sharma, M., Tripathi, M., Prasad, R.: Microbial biofilms: recent advances and progress in environmental bioremediation. Sci. Total Environ. **824**, 153843 (2022)
45. Kumar, V., Prasad, R., Kumar, M.: Rhizobiont in Bioremediation of Hazardous Waste. Springer (2021)
46. Sharma, P.: Efficiency of bacteria and bacterial assisted phytoremediation of heavy metals: an update. Bioresour. Technol. **328**, 124835 (2021)
47. Suganthi, S.H., Murshid, S., Sriram, S., Ramani, K.: Enhanced biodegradation of hydrocarbons in petroleum tank bottom oil sludge and characterization of bio-catalysts and biosurfactants. J. Environ. Manag. **220**, 87–95 (2018)
48. Sarkar, P., Roy, A., Pal, S., Mohapatra, B., Kazy, S.K., Maiti, M.K., Sar, P.: Enrichment and characterization of hydrocarbon-degrading bacteria from petroleum refinery waste as potent bioaugmentation agent for in situ bioremediation. Bioresour. Technol. **242**, 15–27 (2017)
49. Imron, M.F., Kurniawan, S.B., Ismail, N.I., Abdullah, S.R.S.: Future challenges in diesel biodegradation by bacteria isolates: a review. J. Clean. Prod. **251**, 119716 (2020)
50. Xu, X., Liu, W., Tian, S., Wang, W., Qi, Q., Jiang, P., Gao, X., Li, F., Li, H., Yu, H.: Petroleum hydrocarbon-degrading bacteria for the remediation of oil pollution under aerobic conditions: a perspective analysis. Front. Microbiol. **9**, 2885 (2018)
51. Gupta, A., Gupta, R., Singh, R.L.: Microbes and environment. In: Singh, R. (ed.) Principles and Applications of Environmental Biotechnology for a Sustainable Future, pp. 43–84. Springer, Singapore (2017)
52. Roy, A., Dutta, A., Pal, S., Gupta, A., Sarkar, J., Chatterjee, A., Saha, A., Sarkar, P., Sar, P., Kazy, S.K.: Biostimulation and bioaugmentation of native microbial community accelerated bioremediation of oil refinery sludge. Bioresour. Technol. **253**, 22–32 (2018)
53. Mahar, H., Training, T., Jatoi, A.S., Akhtar, K.S., Omer, M., Azeem, S.: Characterization of the sludge deposits in crude oil storage tanks. J. Facul. Eng. Technol. **23**, 57–64 (2016)
54. Sanders, E.R.: Aseptic laboratory techniques: plating methods. J. Vis. Exp. **11**(63), e3064 (2012)
55. Tripathi, N., Sapra, A.: Gram Staining. StatPearls (2020)

Microstructural and Strength Evolutions of Kaolin Geopolymer Based Ceramics at Different Sintering Temperature

Nur Bahijah Mustapa[1], Romisuhani Ahmad[1,2](✉), Mohd Mustafa Al Bakri Abdullah[2], Andrei Victor Sandu[3,4], and Selva Vinayakkan A/L Sures[2]

[1] Faculty of Mechanical Engineering & Technology, Universiti Malaysia Perlis (UniMAP), Perlis, Malaysia

[2] Center of Excellence Geopolymer and Green Technology (CEGeoGTech), Faculty of Chemical Engineering & Technology, Universiti Malaysia Perlis (UniMAP), Perlis, Malaysia

romisuhani@unimap.edu.my

[3] Faculty of Material Science and Engineering, Gheorghe Asachi Technical University of Iasi, Iasi, Romania

[4] Academy of Romanian Scientists (AORS), Bucharest, Romania

Abstract. Ceramic materials generated by combining compacting and sintering have become more essential in the current industry according to the superior mechanical and physical properties. This research aims to understand the effects of different sintering temperatures on kaolin geopolymer based ceramics and explore their microstructure and strength evolution under varying sintering temperature (100 °C, 300 °C, 500 °C, 700 °C, 900 °C and 1100 °C). XRF analysis underscores the dominance of silicon oxide (SiO_2) and aluminum oxide (Al_2O_3) in the geopolymer structure while XRD patterns reveal a dynamic shift in crystalline phases, with nepheline emerging dominantly at higher temperatures. SEM analysis vividly illustrates structural changes, emphasizing increased density and reduced porosity as temperatures rise. Flexural strength peaks at 1100 °C, aligning with heightened density, indicating optimal sintering temperature. This research finding addresses the gap in knowledge concerning the transformation from amorphous to crystalline phases, such as mullite and nepheline, with the overarching goal of unlocking the potential for improved ceramic performance in various industrial applications, including construction materials, refractories, and advanced ceramics.

Keywords: Geopolymer · Geopolymer Based Ceramics · Ceramics · Sintering · Sintering Temperature

A. V. Sandu et al. (Eds.): ICIR EUROINVENT 2024, SPM 64, pp. 81–88, 2025.
https://doi.org/10.1007/978-3-031-81198-2_7

1 Introduction

In recent years, the application of ceramic materials in manufacturing has seen a notable rise. Typically, ceramics are non-metallic and inorganic substances characterized by high melting points, necessitating production and utilization at elevated temperatures. The development of geopolymer ceramics has emerged as a compelling research focus due to the demand for sustainable and eco-friendly materials, as their production requires less energy compared to conventional techniques [1–3].

Geopolymer technology has piqued the interest of industry due to its unique properties designed for specific uses. These geopolymers are formed when aluminosilicates undergo activation in an alkaline setting, resulting in the polymerization of silicon, oxygen, and aluminum elements into amorphous three-dimensional frameworks [4, 5]. Alkaline solutions, predominantly composed of potassium or sodium, not only facilitate geopolymerization to solidify colloids but also stabilize the charge of aluminosilicate tetrahedral structures through their alkali metal cations [6]. Due to their resistance to high temperatures, low permeability, strong adhesive properties, durability, excellent chemical corrosion resistance, and environmental benefits, geopolymers have garnered significant interest [7, 8].

The geopolymerization process of raw materials resulted in the formation of amorphous to sub-crystalline spatial structures.

The process was breakdown into 3 steps: (i) dissolution of aluminosilicates materials and alkaline activator; (ii) gelation; (iii) polymerization [9–12]. The structure of a geopolymer consists of polymeric Si–O–Al framework tetrahedral, with shared oxygen forming 3-D net-work, and metal cations (sodium, potassium, lithium, or calcium) and bound water molecules stabilizing the structure.

Recent advancements in technology have unveiled a new utilization of geopolymer through its conversion into ceramics upon heating. The process involves solid-state sintering, wherein the green body is subjected to high temperatures. This approach of ceramic production using geopolymer precursors has demonstrated enhanced strength compared to traditional methods. Additionally, the sintering temperature significantly influences the strength of the geopolymer ceramic body [13]. Sintering, as a thermally driven adhesion mechanism, enhances particle contact and their subsequent merging. This process results in the closure of open pores, reducing water absorption and enhancing pore strength. The geopolymer, once densely heated, develops a glassy phase, thus acquiring ceramic properties [14].

This study focusing on the transformation from amorphous to crystalline phases of the kaolin geopolymer based ceramics when exposed to different sintering temperature (100 °C, 300 °C, 500 °C, 700 °C, 900 °C and 1100 °C). To elucidate the strength evolution of the ceramics, flexural strength analysis and density were studied.

2 Materials and Methods

2.1 Raw Materials

Kaolin from the Associated Kaolin Industries Sdn. Bhd., Malaysia, served as the main aluminosilicate source for the geopolymer. The chemical composition of the raw kaolin is recorded in Table 1. The main compound of the kaolin is alumina (Al_2O_3) and silica (SiO_2). A mixture of sodium silicate (Na_2SiO_3) and sodium hydroxide (NaOH) solutions were used as an alkali activator during geopolymerization. The NaOH solution was prepared in a volumetric flask and cooled to room temperature for 24 hours at a molarity of 12 M. The sodium hydroxide was provided by Formosoda-P, Taiwan with 99% purity while sodium silicate was sourced from South Pacific Chemicals Industries Sdn. Bhd. (SPCI), Pahang, Malaysia.

2.2 Kaolin Geopolymer Based Preparation

The geopolymer samples were made from kaolin mixed with the mixture of Na_2SiO_3/NaOH at a fixed ratio of 0.24, where the solid-to-liquid ratio is 1.0. Kaolin was mixed with the alkaline activator until it become slurry, before being cured in electrical oven at 80 °C overnight. After hardening, the geopolymer samples underwent milling and sieving to achieve a finely powdered consistency. The resulting green body of the geopolymer was then compacted under a pressure of 4.5 tons for 2 minutes, shaping it into a bar-like structure. These compacted specimens were then subjected to sintering at different temperatures ranging from 100 °C to 1100 °C, employing a constant heating rate of 5 °C/min and a holding time of 180 minutes. This was conducted to investigate the impact of sintering on the properties of the produced ceramics based on kaolin geopolymer.

2.3 Characterization

The microstructure of both unsintered and sintered geopolymer ceramics was examined using a JEOL JSM-6460LA scanning electron microscope (SEM). To prepare for SEM observation, the fracture surfaces of the geopolymer ceramics were coated with gold using a JEOL JFC 1600 model auto fine coater. The phase composition of the samples was assessed using X-ray diffraction (XRD) analysis conducted with a SHIMADZU XRD 6000 instrument, covering a range from 10° to 85°. The samples were ground into a fine powder using a ring mill to ensure optimal penetration of X-rays and generation

Table 1. Chemical composition of raw kaolin.

Element	SiO_2	Al_2O_3	K_2O	Fe_2O_3	TiO_2	MnO_2	ZrO_2
Mass (wt. %)	54.00	31.70	6.05	4.89	1.14	0.11	0.10

of diffraction patterns. Subsequently, the powdered sample was evenly distributed and mounted onto a sample holder. The data collected were analyzed using X'Pert HighScore Plus software. The mechanical properties, particularly the flexural strength, of the geopolymer ceramics were assessed. The three-point bending method, following the protocols outlined in ASTM C-1163b, was utilized, employing a support span length of 30 mm and a crosshead speed of 0.3 mm/min. The density of the geopolymer both before and after the sintering process was determined by applying the principles of Archimedes' theory.

3 Result and Discussion

Figure 1 shows the SEM micrograph of unsintered and sintered kaolin geopolymer at various sintering temperatures of 100 °C, 300 °C, 500 °C, 700 °C, 900 °C, and 1100 °C, imaging at the fracture section. Minimal changes were observed in the geopolymer sintered at 100 °C and 300 °C. However, as the temperature rose to 300 °C and 500 °C, enhancements in surface texture and particle agglomeration within the geopolymer matrix became evident. At 700 °C, densification of the geopolymer matrix commenced alongside the emergence of crystalline phases, notably nepheline, as indicated by XRD analysis. Subsequent temperature increments to 900 °C and 1100 °C further promoted densification, leading to the appearance of closed pores and interconnected pores. These pores were linked to the collapse of the geopolymer matrix induced by phase

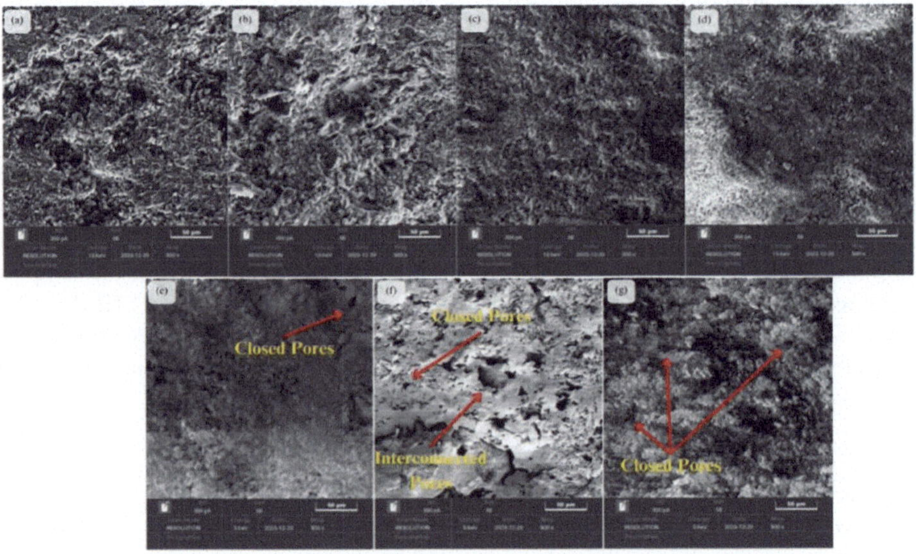

Fig. 1. SEM micrograph of ceramic samples for (a) unsintered and sintered samples at (b) 100 °C, (c) 300 °C, (d) 500 °C, (e) 700 °C, (f) 900 °C, and (g) 1100 °C.

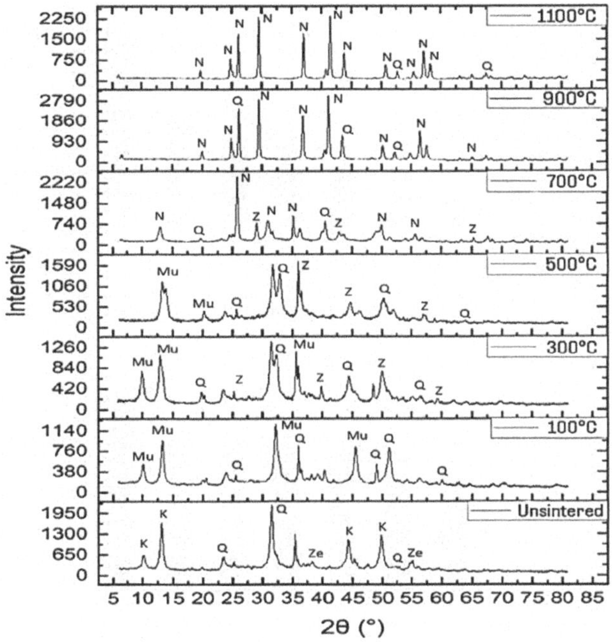

Fig. 2. XRD patterns of unsintered and sintered kaolin geopolymer at different sintering temperature.

transformation and geopolymer decomposition. The quantity and size of the closed pores escalated due to the increased liberation of hydrogen gas [15, 16].

Figure 2 shows the XRD pattern of unsintered and sintered kaolin geopolymer at different sintering temperatures. In the unsintered state, peaks corresponding to quartz, kaolinite, and zeolite were observable within the diffraction angle range of 10°–80° (2θ), attributed to the presence of raw kaolin. The unsintered kaolin geopolymer exhibited the coexistence of kaolinite (K), quartz (Q), and zeolite (Ze). Initiation of the geopolymerization process ensued through the dissolution of kaolin and the alkaline activator. Upon heat treatment, a transition from an amorphous to a crystalline state was evidenced by the alteration in crystallographic composition, characterized by the disappearance of the amorphous phase hump.

With increasing sintering temperatures, intermediate phases such as quartz and muscovite became apparent. At the temperatures of 100 °C, 300 °C, and 500 °C, amorphous humps and reflections from quartz were introduced by the raw kaolin. The emergence of the nepheline crystalline phase was observed at 700 °C, coexisting with residual zircon and quartz. The crystal formations exhibited a dynamic evolution, as the peaks corresponding to nepheline became sharper and more pronounced over time. Sintering temperatures reaching up to 1100 °C facilitated the formation of the mullite phase, contributing to the thermochemical stability properties of the kaolin geopolymer ceramics [17]. Meanwhile, nepheline enhances both the physical and mechanical characteristics of ceramics, including their density and strength, which has been proved by the flexural strength analysis.

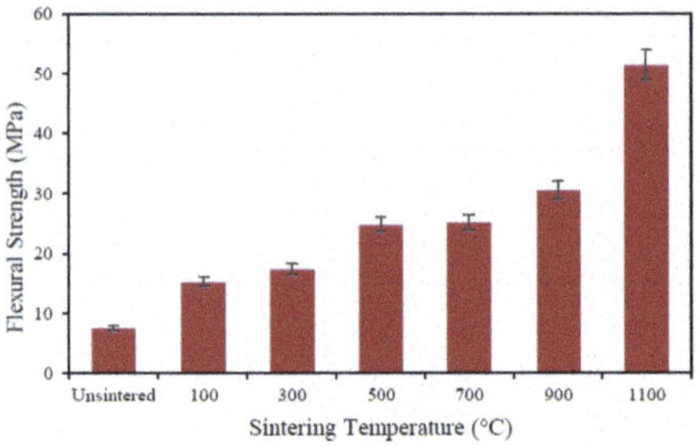

Fig. 3. Flexural strength of unsintered and sintered geopolymer ceramics.

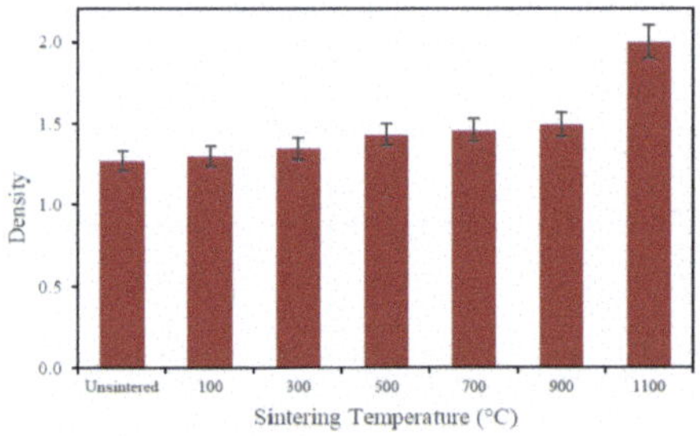

Fig. 4. Density of unsintered and sintered geopolymer ceramics.

Figures 3 and 4 show the flexural strength and the density of the sintered and unsintered geopolymer, respectively.

The flexural strength of kaolin geopolymers ceramics exhibited a rise with increasing the sintering temperature. This increase in strength was attributed to exothermic reactions, wherein the further geopolymerization in the unreacted precursor materials facilitated the formation of more reaction products [18]. The strength of the geopolymer ceramics is highly dependent on the densification and the porosity [19]. The flexural strength reaches its maximum (51.31 MPa) at 1200 °C, which is in align with the density trend obtained. The densification at higher sintering temperatures leads to increased flexural strength and that the combination of polymerization and sintering processes enhances strength. The data from flexural and density validates the SEM and XRD findings, providing an in-depth understanding of the intricate relationship between sintering temperatures and the flexural strength of kaolin geopolymer ceramics.

4 Conclusion

This comprehensive investigation into the fabrication of kaolin geopolymer-based ceramics at varying sintering temperatures has yielded valuable insights into their microstructural evolution and mechanical properties. The research methodically investigated the influence of sintering temperatures spanning from 100 °C to 1100 °C on the microstructure and behavior of the geopolymer ceramics. Utilizing SEM and XRD analyses, the study showed a dynamic evolution in the microstructure, highlighting the development of crystalline phases and trends towards densification, particularly at higher temperatures.

SEM micrographs reveal that as the temperature escalates from 100 °C to 1100 °C, there is a notable enhancement in densification, accompanied by the emergence of a smoother surface texture and a reduction in porosity. XRD patterns illustrate a dynamic progression in crystalline structures, characterized by the gradual sharpening and intensification of nepheline peaks, where it is first appears at 700 °C.

According to mechanical testing, including flexural strength analysis, there was a direct correlation observed between sintering temperature and the resulting increase in strength, which reached its maximum at 1100 °C. Density analysis corroborated these trends, demonstrating the interplay between densification and rising temperatures, emphasizing the role of temperature-induced porosity reduction.

Acknowledgements. The authors of the present work wish to dedicate their great thanks to the Centre of Excellence Geopolymer and Green Technology, (CeGeoGTech), UniMAP and the Faculty of Mechanical Engineering Technology, UniMAP. This research was funded by the Fundamental Research Grant Scheme (FRGS) under grant number FRGS/1/2021/TK0/UNIMAP/02/17 from the Ministry of Higher Education, Malaysia.

References

1. Liu, Y., Liu, L., Mei, H., Liu, Q., Lai, X.: A modified rate-dependent peridynamic model with rotation effect for dynamic mechanical behavior of ceramic materials. Comput. Methods Appl. Mech. Eng. **388**, 114246 (2022)
2. Li, H., Liu, Y., Liu, Y., Zeng, Q., Hu, K., Lu, Z., Liang, J.: Effect of sintering temperature in argon atmosphere on microstructure and properties of 3D printed alumina ceramic cores. J. Adv. Ceram. **9**, 220–231 (2020)
3. Almutairi, A.L., Tayeh, B.A., Adesina, A., Isleem, H.F., Zeyad, A.M.: Potential applications of geopolymer concrete in construction: a review. Case Stud. Construct. Mater. **15**, e00733 (2021)
4. Moutaoukil, G., Sobrados, I., Alehyen, S., Taibi, M.H.: Monitoring the geopolymerization reaction of geopolymer foams using 29Si and 27Al MAS NMR. Fortschr. Mineral. **14**(5), 516 (2024)
5. Luhar, S., Nicolaides, D., Luhar, I.: Fire resistance behaviour of geopolymer concrete: an overview. Buildings. **11**(3), 82 (2021)
6. Chen, W., Garofalo, A.C., Geng, H., Liu, Y., Wang, D., Li, Q.: Effect of high temperature heating on the microstructure and performance of cesium-based geopolymer reinforced by cordierite. Cem. Concr. Compos. **129**, 104474 (2022)
7. Wu, Y., Lu, B., Bai, T., Wang, H., Du, F., Zhang, Y., Cai, L., Jiang, C., Wang, W.: Geopolymer, green alkali activated cementitious material: synthesis, applications and challenges. Constr. Build. Mater. **224**, 930–949 (2019)
8. Ji, Z., Li, M., Su, L., Pei, Y.: Porosity, mechanical strength and structure of waste-based geopolymer foams by different stabilizing agents. Constr. Build. Mater. **258**, 119555 (2020)

9. Yuan, J., Li, L., He, P., Chen, Z., Lao, C., Jia, D., Zhou, Y.: Effects of kinds of alkali-activated ions on geopolymerization process of geopolymer cement pastes. Constr. Build. Mater. **293**, 123536 (2021)
10. Hsieh, Y.C., Lee, W.H., Liao, P.H.: Using geopolymer technology on synthesizing leucite ceramics. Polymers. **13**(21), 3621 (2021)
11. Ahmed, H.U., Mahmood, L.J., Muhammad, M.A., Faraj, R.H., Qaidi, S.M., Sor, N.H., Mohammed, A.S., Mohammed, A.A.: Geopolymer concrete as a cleaner construction material: an overview on materials and structural performances. Clean. Mater. **5**, 100111 (2022)
12. Zribi, M., Baklouti, S.: Investigation of phosphate based geopolymers formation mechanism. J. Non-Cryst. Solids. **562**, 120777 (2021)
13. Jaya, N.A., Al Bakri Abdullah, M.M., Ghazali, C.M.R., Binhussain, M., Kamarudin, H., Ahmad, R., Ekaputri, J.J.: Correlation between Na2SiO3/NaOH and NaOH molarity to flexural strength of geopolymer ceramic. Appl. Mech. Mater. **754**, 152–156 (2015)
14. Ramli, M.I.I., Salleh, M.A.A.M., Abdullah, M.M.A.B., Aziz, I.H., Ying, T.C., Shahedan, N.F., Kockelmann, W., Fedrigo, A., Sandu, A.V., Vizureanu, P., Chaiprapa, J.: The influence of sintering temperature on the pore structure of an alkali-activated kaolin-based geopolymer ceramic. Materials. **15**(7), 2667 (2022)
15. Sawan, S.A., Zawrah, M.F., Khattab, R.M., Abdel-Shafi, A.A.: In-situ formation of geopolymer foams through addition of silica fume. Preparation and sinterability. Mater. Chem. Phys. **239**, 121998 (2020)
16. Aziz, I.H., Al Bakri Abdullah, M.M., Salleh, M.A.A.M., Yoriya, S., Razak, R.A., Mohamed, R., Baltatu, M.S.: The investigation of ground granulated blast furnace slag geopolymer at high temperature by using electron backscatter diffraction analysis. Arch. Metall. Mater. **67**, 227–231 (2022)
17. Deutou, J.G.N., Kaze, R.C., Kamseu, E., Sglavo, V.M.: Controlling the thermal stability of kyanite-based refractory geopolymers. Materials. **14**(11), 2903 (2021)
18. Yong-Sing, N., Yun-Ming, L., Cheng-Yong, H., Abdullah, M.M.A.B., Pakawanit, P., Chan, L.W.L., Hui-Teng, N., Shee-Ween, O., Wan-En, O., Yong-Jie, H.: Thin fly ash/ladle furnace slag geopolymer: effect of elevated temperature exposure on flexural properties and morphological characteristics. Ceram. Int. **48**(12), 16562–16575 (2022)
19. Zeng, L., Sun, H.J., Peng, T.J., Zheng, W.M.: The sintering kinetics and properties of sintered glass-ceramics from coal fly ash of different particle size. Result Phys. **15**, 102774 (2019)

Effect of Sintering Temperature on Structural, Physical and Mechanical Properties of The Activated-HAp Ceramic

Wan Mohd Arif Wan Ibrahim[1,2] (✉), Mohd Mustafa Al Bakri Abdullah[2] (✉), Noorina Hidayu Jamil[3], Hasmaliza Mohamad[4], Mohd Arif Anuar Mohd Salleh[2], and Petrică Vizureanu[5]

[1] Center of Excellence Geopolymer and Green Technology (CEGeoGTech), Universiti Malaysia Perlis (UniMAP), Arau, Perlis, Malaysia

[2] Faculty of Chemical Engineering & Technology, Universiti Malaysia Perlis (UniMAP), Arau, Perlis, Malaysia

wmarif@unimap.edu.my; mustafa_albakri@unimap.edu.my

[3] Faculty of Mechanical Engineering & Technology, Universiti Malaysia Perlis (UniMAP), Arau, Perlis, Malaysia

[4] School of Materials and Mineral Resources Engineering, Universiti Sains Malaysia, Gelugor, Penang, Malaysia

[5] Faculty of Material Science and Engineering, Gheorghe Asachi Technical University of Iasi, Iasi, Romania

Abstract. This work investigates the effect of varied sintering temperatures on the structural, physical, and mechanical properties of activated-hydroxyapatite (HAp) ceramic, with a focus on its potential implications for biomedical applications. Activated metakaolin (MK)/HAp specimens were prepared using an alkaline activation technique, maintaining a fixed metakaolin loading in HAp at a liquid-to-solid (L/S) ratio of 1.25. These specimens were subjected to pressing and sintering across a temperature range of 800 °C to 1200 °C. Results were comprehensively analysed through phase analysis, thermal analysis, surface morphology study, and evaluations of mechanical and physical properties. Phase analysis revealed the formation of multiple crystalline phases within the sintering temperature range of 800 °C to 1100 °C, while sintering at 900 °C resulted in the formation of a single monoclinic hydroxyapatite phase, providing the highest recorded tensile strength of 12.52 MPa. X-ray diffraction (XRD) results also indicated the presence of a semi-crystalline phase at 1200 °C, consistent with crystallization peak temperatures (Tp) of 1070 °C observed in thermal analysis. In conclusion, this study underscores the significant influence of sintering temperature on the phase stability and overall properties of activated-HAp ceramic, with sintering at 900 °C facilitating the formation of a single monoclinic hydroxyapatite phase. Additionally, the observed low-fusion phenomenon at 1200 °C presents intriguing possibilities for broader biomedical applications.

Keywords: Activated-hydroxyapatite (HAp) · Monoclinic · Low-fusion · Hydroxyapatite phase · Diametrical tensile strength

A. V. Sandu et al. (Eds.): ICIR EUROINVENT 2024, SPM 64, pp. 89–99, 2025.
https://doi.org/10.1007/978-3-031-81198-2_8

1 Introduction

Hydroxyapatite (HAp) is a mineral form of calcium apatite known for its applications in bone grafts and dental implants due to its biocompatibility and similarity to human bone mineral composition [1, 2]. However, the brittleness and limited mechanical strength of pure HAp hinder its use in load-bearing applications. Recent research has focused on enhancing HAp's properties through techniques like incorporating secondary materials and optimizing sintering conditions, such as temperature [3].

Sintering is a crucial process in ceramic manufacturing that involves consolidating powders at high temperatures, leading to significant impacts on the microstructure and properties of the material. Recent studies have focused the effect of sintering temperature on the phase composition and mechanical characteristics of HAP, thereby affecting its suitability for various applications [4]. For instance, the sintering temperature has been demonstrated to play a significant process in the microstructure evolution of nano-hydroxyapatite powder for bone tissue engineering. Additionally, the incorporation of TiH_2 in hydroxyapatite has been investigated to understand its effect on the microstructure at different sintering temperatures. Furthermore, the role of $MgAl_2O_4$ powder packing on the phase stability of hydroxyapatite during sintering has been explored, demonstrating the importance of additional materials in the sintering process. By investigating the impact of sintering temperature on activated-HAp ceramic properties, researchers aim to optimize conditions for composite materials, focusing on phase transitions, thermal behavior, surface morphology, and mechanical strength within the temperature range of 800 °C to 1200 °C [3].

Considering the alkaline activation of HAp prior to the sintering process could be beneficial. Previous studies have demonstrated that initiating geopolymerization in sintered kaolin ceramics at 900 °C results in excellent compressive strength [5]. Additionally, using kaolin [6] as a source of alkaline-activated materials (AAM) has shown good bioactivity of the specimens [7, 8]. Previous research has demonstrated that sintering temperature plays an important role in shaping the microstructural, mechanical, and biological properties of HAp. By conducting a comprehensive analysis of how sintering temperatures affect activated-HAp specimens, particularly in terms of crystalline phase formation and implications for biomedical use, this study aims to provide valuable insights into the sintering behavior of activated-HAp ceramic. The findings are expected to contribute to the development of advanced HAp-based materials with improved performance for diverse biomedical applications, enhancing their efficacy and reliability in clinical settings.

2 Materials and Methods

2.1 Materials

The following materials were utilized in the formulation of the alkaline-activated MK/HAp (activated-HAp) ceramic: hydroxyapatite (HAp), kaolin, sodium hydroxide (NaOH), sodium silicate (Na_2SiO_3) pellets, and water. High-purity HAp powder

Table 1. Chemical composition of HAp and metakaolin.

Chemical Composition (Compound)	CaO	P_2O_5	Al_2O_3	SiO_2	Na_2O	TiO_2	Fe_2O_3	MgO	K_2O
HAp	61.35	18.63	14.00	0.38	0.10	–	–	5.20	–
MK	–	–	38.70	59.00	0.40	1.10	0.60	–	0.10

(heavy metals ≤20.0 ppm) with a particle size of 10 μm was obtained from Sigma-Aldrich (USA), and free-flow grade kaolin powder was also sourced from Sigma-Aldrich (USA). Both materials were utilized as received.

The chemical compositions of HAp and kaolin were determined by X-ray fluorescence (XRF) analysis, as shown in Table 1. Metakaolin (MK), the primary precursor for geopolymerization, was produced by calcining raw kaolin at 750 °C for three hours in an electric oven. The NaOH used, Formosa-P type with 99.0% purity, was sourced from Formosa Plastic Corporation (Taipei, Taiwan). Additionally, Na_2SiO_3 was obtained from South Pacific Chemicals Industries Sdn. Bhd. (SPCI, Selangor, Malaysia).

2.2 Preparation of Activated MK-HAp Ceramic

MK/HAp was activated using geoplymerization technique by mixing 30 g of metakaolin with 100 g of HAp and subjecting them to ball-milling at 200 rpm for 3 hours in PTFE mill jar with ZrO_2 balls. The geopolymerization process involved mixing of sodium silicate solution with 8 M of NaOH in a 1:2 ratio. The mixture was kept at room temperature for 24 hours prior to ceramic activation. The activated MK/HAp in the form of paste was dried at 60 °C for 48 hours, crushed, dry milled, sieved (< 63 μm), and compacted at 40 MPa to form green compacts. The green compacts were then sintered in a two-steps process: first heating to 500 °C at 3 °C/min for two hours, then heating at again 3 °C/min to a temperature range of 800–1200 °C for another two hours to prepared the activated MK/HAp ceramic specimens.

2.3 Characterizations and Testing

Prior to characterization and testing the sintered sample was physically observed to look at the physical transformation of the specimens over the different sintering temperatures. The phase and crystalline structure of the specimens was obtained by a Bruker D2 Phaser X-Ray diffractometer (XRD) (Bruker, Germany) with Cu Kα radiation (30 mA and 40 kV). The thermal properties of activated-HAp were simultaneously studied by differential scanning calorimetry (TGA/DSC 2 HT Mettler Toledo, Columbus, OH, USA) up to the temperature of 1200 °C. Vega TESCA scanning electron microscope (SEM) (Tescan, Brno, Czech Republic) was used to study the microstructure of the as-sintered specimens. Tensile strength test was conducted on the sintered specimens, which were shaped into cylindrical pallets of 15 mm in diameter and 5 mm in height in accordance with ASTM C 496 standards. Diametral tensile strength (DTS) values were measured for three specimens from each mix design using a Shimadzu, AG-X Plus universal testing machine (Shimadzu, Kyoto, Japan) at a cross-head speed of 0.600 mm/min. The DTS values were calculated using Eq. (1):

$$\sigma = \frac{2F}{\pi DT} \tag{1}$$

where:

D = diameter of the specimen [mm],
T = thickness of the specimen [mm],
F = force causing the destruction of the specimen [N],
σ = diametral tensile strength (DTS) [MPa].

The results were obtained using the mean value ± standard deviation of at least three independent specimens [9].

3 Results and Discussion

3.1 Physical Observation—*Visual Observation*

Figure 1 shows the visual observations of the activated-HAp and pure HAp specimens sintered at various temperatures reveal distinct differences in their surface characteristics, densification behaviour, and physical dimensions.

At 800 °C, both pure and activated-HAp specimens exhibit rough and porous surfaces, indicating incomplete sintering. The activated-HAp specimens appear slightly smoother than the pure HAp, suggesting some improvement in densification. As the sintering temperature increases to 900 °C, the pure HAp specimens show reduced porosity and increased smoothness, indicating a higher degree of sintering. The activated-HAp specimens at this temperature exhibit a more uniform and dense structure with significant reductions in surface irregularities and porosity, demonstrating enhanced densification [3]. The activated-HAp specimens appear highly uniform

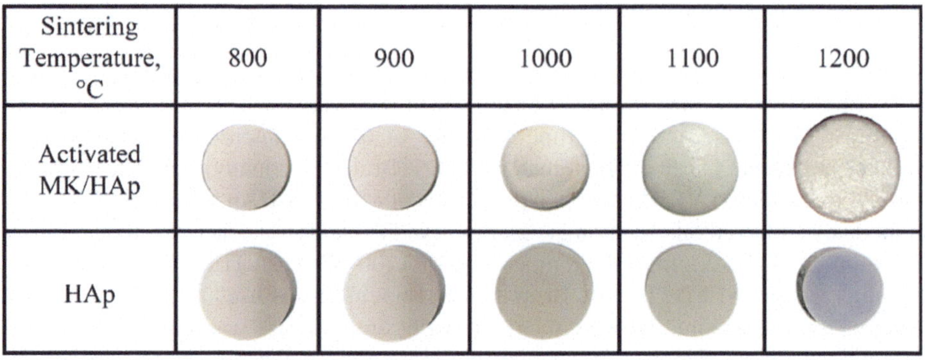

Fig. 1. Visual observation of activated-HAp and pure HAP specimens sintered at different temperatures.

and dense, suggesting optimal sintering with minimal remaining porosity. Notably, the physical dimensions of the specimens are consistent, allowing for direct comparison and highlighting the effectiveness of the sintering process at different temperatures.

At 1000 °C, both pure and activated-HAp specimens continue to densify, with the pure HAp showing further reduced porosity and a more compact structure. At 1100 °C, the pure HAp specimens present a smooth and dense surface with minimal observable pores, while the activated-HAp specimens started to fuse where a glassy surface is visible. At 1200 °C, activated-HAp specimens start to exhibit a glassy appearance, suggesting over-sintering and the onset of phase transformation. The specimen show a fused, glassy surface and expended, indicating excessive sintering and potential structural collapse. Figure 2 shows the thermal analysis of differential scanning calorimeter (DSC) of the specimen. The exothermic peak associated with the crystallization of glass (T_p) was obtained at 1068 °C and it is promoting fusion and compromising the structural integrity of the specimen. It is also observed that pure HAp specimens exhibit a characteristic blueish tint at 1200 °C, which is common and indicative of phase changes and high densification [3].

The physical dimensions of all specimens remain comparable, emphasizing that the observed changes are due to microstructural modifications rather than size variations. These observations suggest that the optimal sintering temperature for achieving balanced mechanical strength and structural integrity in activated MK/HAp ceramics is around 900 °C, whereas for pure HAp, the optimal temperature is around 1200–1250 °C, where it achieves maximum densification and mechanical properties.

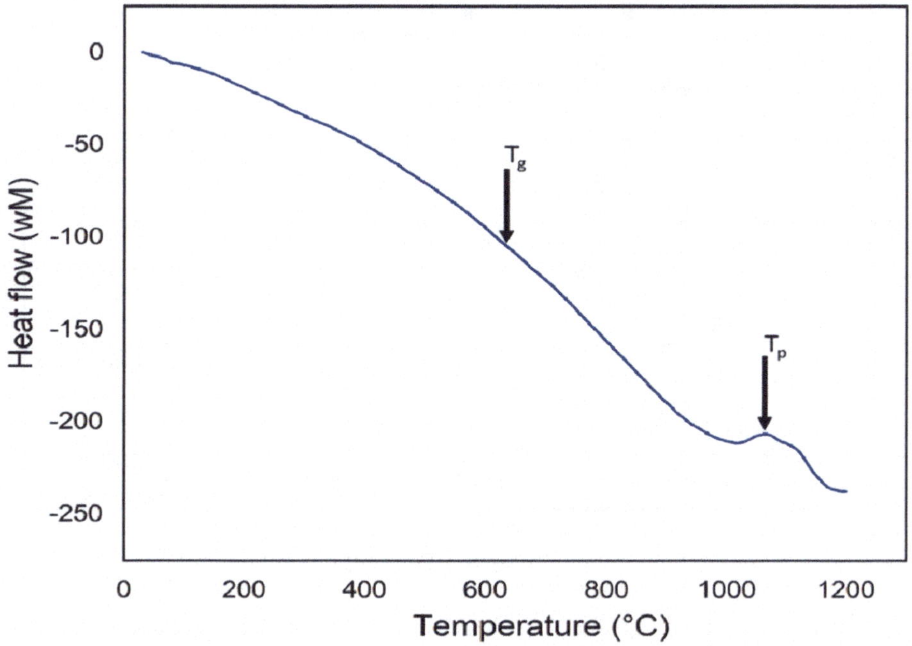

Fig. 2. DSC curve of activated-Hap.

3.2 Phase Stability

The X-ray diffraction spectrum of the specimens sintered at different temperatures 800, 900, 1000, 1100 and 1200 °C is depicted in Fig. 3. The patterns reveal the presence of well crystallized phases in all specimens, except for specimens sintered at 1200 °C. Phase transition is clearly observable across the different sintering, indicating a direct correlation with the material's stability.

Sintering at 900 °C results in the formation of the monoclinic HAp phase, identified by peaks at 23.49°, 27.69°, 33.6.12°, 46.98°, and 68.28° 2θ (Powder Diffraction File-PDF No. 01-089-4405). To better undertand the formation of the monoclinic HAp, and considering that the XRD characteristic peaks (at dominant peaks) of this phase overlap with those of the hexagonal phase due to geometric similarity, thorough observations were made on the peaks at higher angles (2θ). A details discussion on confirming the monoclinic structure has been provided in a previous study [10]. The monoclinic phase is thermodynamically more stable and mechanically superior, making it more suitable for various medical applications. A similar observation was reported by Morgan et al. who revealed the monoclinic HAp phase alongside the hexagonal phase specimen sintered at 900 °C [11].

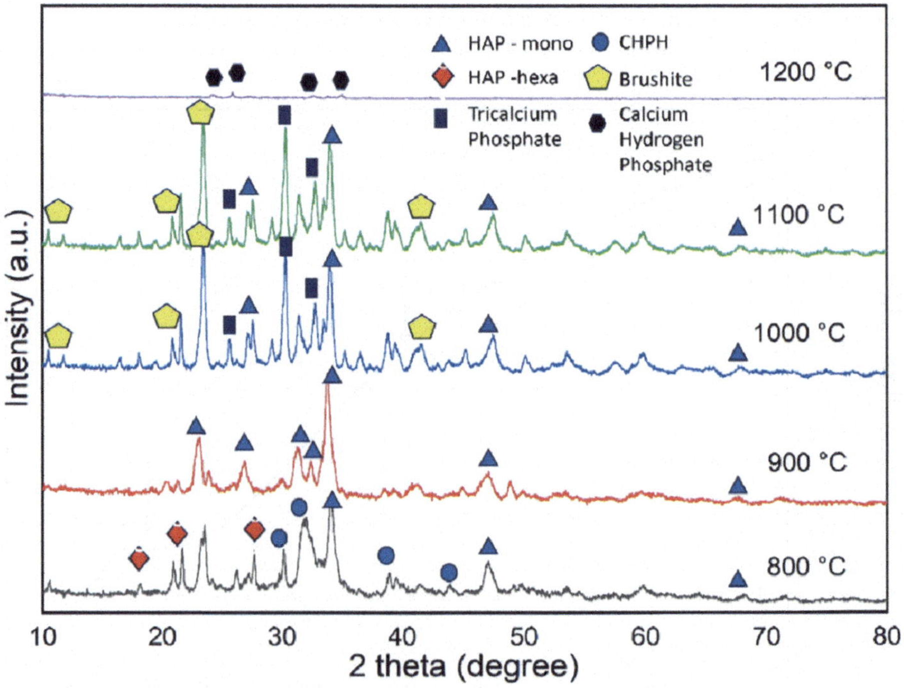

Fig. 3. XRD spectra of activated-HAp sintered at different sintering temperature.

X-ray patterns of the activated-HAp at 800, 1000 and 1100 °C, reveal the presence of multiple phases included HAp, decomposition phases of HAp and others. The decomposition characteristics of HAp are reported variedly in the literature. It was reported that in temperature range from 600 to 1200 °C HAp decomposes into β-tricalcium phosphate [4]. Structural analysis at 1000 and 1100 °C shows insignificant phase transition, where both contained crystalline phases of hydroxyapatite (PDF card number: 01-089-4405) and brushite (PDF card number: 00-009-0077) as well as tricalcium phosphate (PDF card number: 00-048-0488).

The sintering temperature at 1200 °C shows a diffuse halo peak with a few visible peaks at low intensity representing a semi-crystalline structure. The peaks indicate the existence of low intensity calcium hydrogen phosphate structure (PDF card number: 01-089-3748). The phase commonly serves as a precursor material for hydroxyapatite [12]. Chao mentioned during sintering, substituting of calcium with sodium can induce structural changes and impacted the material's final properties [13].

3.3 Mechanical Analysis

The tensile strength analysis of the sintered activated-HAp ceramic at different sintering temperatures, as depicted in the Fig. 4, reveals significant insights into the material properties. The lowest strength is observed in the pure hydroxyapatite (sintered at 900 °C), which is included as a control.

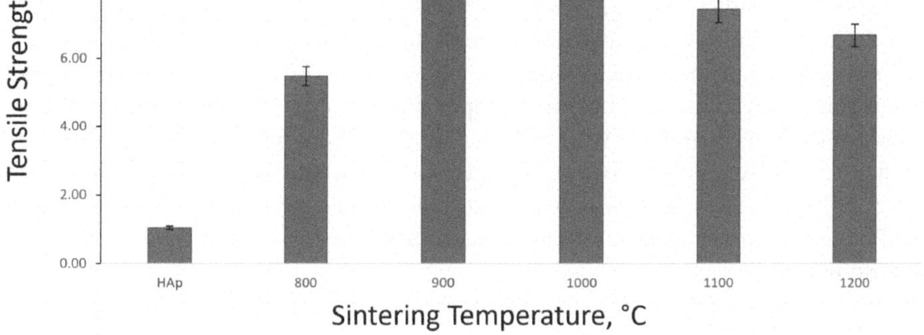

Fig. 4. Tensile strength of activated-HAp sintered at different temperatures.

Previous study showed that sinter at 900 °C results in the high tensile strength for sintered geopolymer-ceramic [5]. This is attributed to the thermo-stable nature of the monoclinic structure, which maintains structural integrity and enhances strength. The optimal sintering temperature for hydroxyapatite composites is around 900 °C, supporting the observed results.

The addition of metakaolin has been found to significantly improve the mechanical properties of the specimen, as evidenced by several researchers [5]. At temperatures up to 1000 °C, the improvement in tensile strength is still observed, although not as pronounced as at 900 °C. This improvement is attributed to the continued stability of the monoclinic structure and the beneficial effects of metakaolin. However, a notable decrease in tensile strength is observed at 1200 °C. This reduction in strength is due to the transformation of the material into a glassy phase, which negatively impacts its mechanical properties. Other finding confirmed that the formation of a glassy phase at high temperatures leads to a reduction in crystallinity and mechanical strength of the material [14].

Overall, the analysis indicates that the optimal sintering temperature for achieving maximum tensile strength in activated-HAp ceramic is around 900 °C. Higher temperatures, particularly 1200 °C, should be avoided due to the adverse effects of phase transformation.

3.4 Morphology

The tensile strength results, as shown in Fig. 4, provide critical insights into the correlation between sintering temperature, microstructure, and mechanical properties of hydroxyapatite ceramics.

The results demonstrate that the tensile strength increases significantly from 800 °C to 900 °C, reaching the highest value at 900 °C. This correlates with microstructural observations where 900 °C shows a significant reduction in porosity and better grain coalescence, leading to enhanced mechanical properties. The activation of HAp by an alkaline technique prior to sintering further aids in achieving an optimal microstructure at a relatively lower temperature of 900 °C, Fig. 5.

At 1000 °C and 1100 °C, the tensile strength decreases slightly but remains relatively high. The microstructure at these temperatures is denser with fewer voids, contributing to good mechanical strength. However, at 1200 °C, the tensile strength drops significantly, Fig. 6. The micrograph (Fig. 6b) at this temperature shows a collapse in the microstructure with no observable micropores, indicating the transformation to a semi-crystalline phase. This is supported by the formation of a glassy surface and the DSC result, which indicate an exothermic peak associated with the crystallization of glass at around 1068 °C, promoting fusion and compromising the structural integrity.

These findings suggest that the optimal sintering temperature for achieving a balance between density, mechanical strength, and structural stability in HAp ceramics is 900 °C.

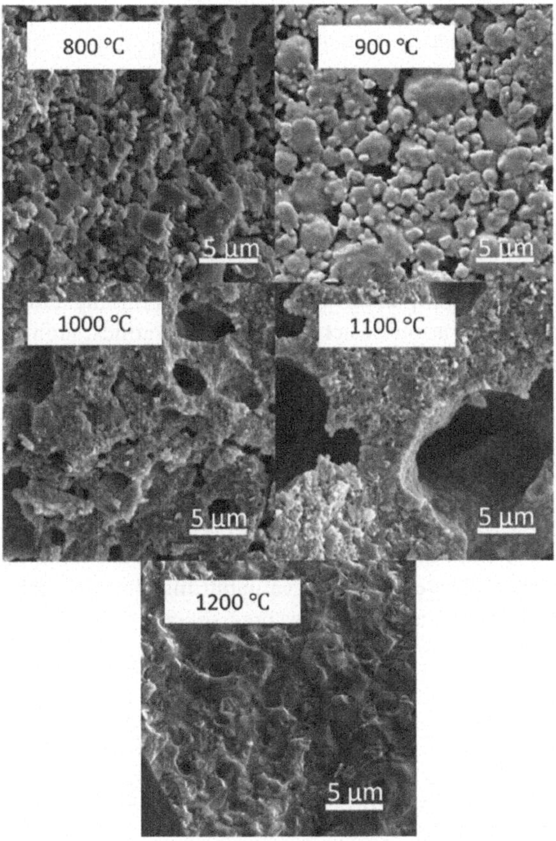

Fig. 5. SEM micrograph of sintered activated-HAp at different temperatures.

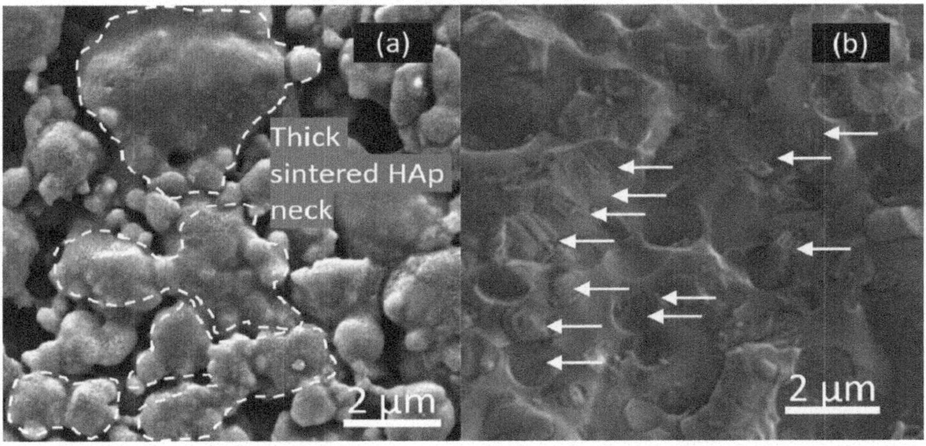

Fig. 6. SEM micrograph (a) of the activated-HAp sintered at 900 °C reveals the thick-necking of the microstructure, and (b) sintered at 1200 °C shows no visible micro-pores due to structure collapse during glass transition.

The inclusion of metakaolin and the initial alkaline activation process significantly contribute to achieving superior properties at this temperature. Further increasing the sintering temperature leads to a degradation in properties due to excessive grain growth and phase transformation.

4 Conclusion

Based on the analysis and comparison with established findings, the sintering temperature is crucial in the preparation of activated-HAp ceramics, significantly influencing their microstructure, phase stability, and mechanical properties. This study successfully identified the optimal sintering temperature for activated-HAp ceramics at 900 °C, corroborated by morphological studies showing a highly uniform and dense structure with minimal porosity at this temperature. The phase stability of activated-HAp is highly dependent on the sintering temperature, with proper sintering enhancing densification and mechanical properties while preventing adverse phase transformations. Additionally, the low-fusion phenomenon observed at 1200 °C, where the material exhibits a glassy surface and structural collapse, provides valuable insights for other medical applications, such as dental prosthetics, where precise control of material properties is crucial.

Acknowledgments. The authors gratefully acknowledge the financial support of the Centre of Excellence Geopolymer and Green Technology (CEGeoGTech), UniMAP. The authors also would like to extend their gratitude to the Ministry of Higher Education for sponsoring the Fundamental Research Grant Scheme (FRGS) FRGS/1/2020/TK0/UNIMAP/01/2.

References

1. Tavoni, M., Dapporto, M., Tampieri, A., Sprio, S.: Bioactive calcium phosphate-based composites for bone regeneration. J. Compos. Sci. **5**(9), 227 (2021)
2. Yazdanpanah, Z., Bahrololoom, M.E., Hadhemi, B.: The effect of sintering temperature on microstructure of glass-reinforced hydroxyapatite biocomposites immersed in simulated body fluid. Ceram. Silik. **50**(4), 298–304 (2015)
3. Trzaskowska, M., Vivcharenko, V., Przekora, A.: The impact of hydroxyapatite sintering temperature on its microstructural, mechanical, and biological properties. Int. J. Mol. Sci. **24**(6), 5083 (2023)
4. Amirthalingam, N., Panchatcharam, S., Deivarajan, T., Paramasivam, M.: Effect of TiH_2 on microstructure of hydroxyapatite sintered at various temperatures. Process. Appl. Ceram. **16**(2), 125–133 (2022)
5. Jamil, N.H., Abdullah, M.M.A.B., Che Pa, F., Mohamad, H., Ibrahim, W.M.A.W., Chaiprapa, J.: Influences of SiO_2, Al_2O_3, CaO and MgO in phase transformation of sintered kaolin-ground granulated blast furnace slag geopolymer. J. Mater. Res. Technol. **9**(6), 14922–14932 (2020)

6. Jamil, N.H., Abdullah, M.M.A.B., Che Pa, F., Mohamad, H., Ibrahim, W.M.A.W., Amonpattaratkit, P., Gondro, J., Sochacki, W., Ibrahim, N.: Self-fluxing mechanism in geopolymerization for low-sintering temperature of ceramic. Materials. **14**(6), 1325 (2021)
7. Sunendar, B., Fathina, A., Harmaji, A., Mardhian, D.F., Asri, L., Widodo, H.B.: The effect of CHA-doped Sr addition to the mechanical strength of metakaolin dental implant geopolymer composite. AIP Conf. Proc. 1887. **1887**, 020020 (2017)
8. Catauro, M., Bollino, F., Papale, F., Lamanna, G.: Investigation of the sample preparation and curing treatment effects on mechanical properties and bioactivity of silica rich metakaolin geopolymer. Mater. Sci. Eng. C. **36**(1), 20–24 (2014)
9. Jurgelane, I., Buss, A., Putnina, M., Loca, D., Locs, J.: Effect of sintering temperature on sorption properties and compressive strength of calcium phosphate ceramic granules. Mater Lett. **282**, 128858 (2021)
10. Ibrahim, W.M.A.W., Abdullah, M.M.A.B., Jamil, N.H., Mohamad, H., Salleh, M.A.A.M., Sandu, A.V., Vizureanu, P., Baltatu, M.S., Sukmak, P.: Alkaline-activation technique to produce low-temperature sintering activated-HAp ceramic. Appl. Sci. (Switzerland). **13**(4), 2643 (2023)
11. Morgan, H., Wilson, R.M., Elliott, J.C., Dowker, S.E.P., Anderson, P.: Preparation and characterisation of monoclinic hydroxyapatite and its precipitated carbonate apatite intermediate. Biomaterials. **21**(6), 617–627 (2000)
12. Hagio, T., Yamauchi, K., Iwai, K.: Magnetically induced crystal alignment of calcium hydrogen phosphate dihydrate. Journal of Materials Science Research. **3**(2), 38–43 (2014)
13. Cho, J.S., Um, S.H., Yoo, D.S., Chung, Y.C., Chung, S.H., Lee, J.C., Rhee, S.H.: Enhanced osteoconductivity of sodium-substituted hydroxyapatite by system instability. J Biomed Mater Res B Appl Biomater. **102**(5), 1046–1062 (2014)
14. Tullu, A.M., Terfasa, T.T., Zerfe, E.A., Tadese, M., Bayene, E., Abebe, A.M., Adoshe, D.M.: Effect of cullet on firing temperature and dielectric properties of porcelain insulator. Heliyon. **8**(2), e08922 (2022)

Author Index

A. V. Sandu et al. (Eds.): ICIR EUROINVENT 2024, SPM 64, p. 101, 2025.
https://doi.org/10.1007/978-3-031-81198-2

The manufacturer's authorised representative in the EU is Springer Nature Customer Service Centre GmbH, Europaplatz 3, 69115 Heidelberg, Germany. If you have any concerns regarding our products, please contact ProductSafety@springernature.com

Printed and bound by CPI Group (UK) Ltd, Croydon, CR0 4YY
01/10/2026
02225977-0002